International Workshop "Nuclear Reactions on Nucleons and Nuclei" (NRNN 2017)

Journal of Physics: Conference Series Volume 1014

Messina, Italy
25 - 26 October 2017

Editors:

Giuseppe Mandaglio
Antonio Trifiro
Marina Trimarchi

ISBN: 978-1-5108-6290-6
ISSN: 1742-6588

International Workshop "Nuclear Reactions on Nucleons and Nuclei" (NRNN 2017)

Journal of Physics: Conference Series Volume 1014

Messina, Italy
25 - 26 October 2017

IOP
Institute of Physics

Printed from e-media with permission by:

Curran Associates, Inc.
57 Morehouse Lane
Red Hook, NY 12571

Some format issues inherent in the e-media version may also appear in this print version.

Printed by Curran Associates, Inc. (2018)

For permission requests, please contact the Institute of Physics
at the address below.

Institute of Physics
Dirac House, Temple Back
Bristol BS1 6BE UK

Phone: 44 1 17 929 7481
Fax: 44 1 17 920 0979

techtracking@iop.org

Additional copies of this publication are available from:

Curran Associates, Inc.
57 Morehouse Lane
Red Hook, NY 12571 USA
Phone: 845-758-0400
Fax: 845-758-2633
Email: curran@proceedings.com
Web: www.proceedings.com

Table of contents

Volume 1014

International Workshop "Nuclear Reactions on Nucleons and Nuclei"

25–26 October 2017, Messina, Italy

Accepted papers received: 20 April 2018
Published online: 10 May 2018

Preface

Papers

Preface

This volume of *Journal of Physics: Conference Series* collects the papers resulting of the invited contributions presented and discussed during the International Workshop "Nuclear Reactions on Nucleons and Nuclei" held in Messina (Italy) during 25 and 26 October, 2017. The scientific event was organized by the University of Messina and it took place in the fascinating and suggestive library room of the "Accademia Peoloritana dei Pericolanti".

Library room of the "Accademia Peoloritana dei Pericolanti",
picture taken during a workshop talk.

During the workshop have been presented more than twenty oral presentation on the following topics:

- **Nuclear reaction dynamics**
- **Heavy and superheavy elements synthesis**
- **Multifragmentation reactions**
- **Nuclear reactions induced by radioactive beams**
- **Exotic nuclei and nuclear reactions of astrophysical interest**
- **Properties of nuclear matter**

have been discussed review, works in progress and future project in this fascinating field of theoretical and experimental physics.
We thank all participants for the scientific presentation, for the interesting discussions

arose after each presented talk, and for their effort to prepare good papers leaving a useful trace of this scientific event. We also thank the referees of the papers for their suggestions and again the authors for their prompt revision of the articles.

We would like to thank the staff of the "Accademia Peloritana dei Pericolanti" for their assistance during the worshop, and the students of University of Messina for their active participation to the scientific discussions.

Some days before the workshop opening, unfortunately, we learned about the pass away of Professor Adam Sobiczewski, a great scientist who left a huge contribution on the nuclear structure knowledge, an honest man and a good friend who inspired with his gentle manners and his deep knowledge of nuclear physics generations of young researchers.

We wish to dedicate this volume in lovely memory of Professor Adam Sobiczewski.

The Editors of the NRNN2017 Proceedings:

Giuseppe Mandaglio (University of Messina and INFN Sez. Catania)
Antonio Trifirò (University of Messina and INFN Sez. Catania)
Marina Trimarchi (University of Messina and INFN Sez. Catania)

Messina, 15rd of April 2018

IOP Conf. Series: Journal of Physics: Conf. Series **1014** (2018) 011001 doi:10.1088/1742-6596/1014/1/011001

IOP Conf. Series: Journal of Physics: Conf. Series **1014** (2018) 011002 doi:10.1088/1742-6596/1014/1/011002

International Workshop

Nuclear Reactions on Nucleons and Nuclei

Messina (Italy), October 25-26, 2017

Editors:

Giuseppe Mandaglio, Marina Trimarchi, Antonio Trifirò

Organizing Committee:

Chairman: Giuseppe Mandaglio (Messina)
Chairman: Antonio Trifirò (Messina)
Chairman: Marina Trimarchi(Messina)
Honorary Chairman: Giorgio Giardina(Messina)

Scientific Secretaries: A. Anastasi (Messina), L. Quattrocchi (Catania)

Organizing Institutions: University of Messina

International Committee:

- G. Adamian (Dubna – Russia)
- N.V. Antonenko (Dubna – Russia)
- B. B. Back (ORNL, USA)
- A. Badalà (INFN- Catania, Italy)
- M. Battaglieri (INFN- Genova, Italy)
- G. Cardella (INFN- Catania, Italy)
- S. Heintz (GSI – Germany)
- G. Giardina (University of Messina, Italy)
- Yu-Gang Ma (Shangai, China)
- G. Mandaglio (University of Messina, Italy)
- Jie Meng (Beijing, China)
- E. Kozulin (JINR, Russia)
- D. Moricciani (INFN Sezione Roma 2, Italy)
- A. K. Nasirov (Dubna – Russia)
- K. Nishio (JAEA, Japan)
- Yu. Ts. Oganessian (JINR, Russia)
- A. Pagano (INFN- Catania, Italy)

IOP Conf. Series: Journal of Physics: Conf. Series **1014** (2018) 011002 doi:10.1088/1742-6596/1014/1/011002

- P. Pedroni (INFN – Pavia, Italy)
- K. Pomorski (Lublin, Poland)
- E. Prasad (Kerala, India)
- F. Riggi (University of Catania)
- F. Rizzo (University of Catania and LNS Catania, Italy)
- A. Szczepaniak (Indiana University, USA)
- A. Sobcziewski (Warsaw – Poland)
- A.Trifirò (University of Messina, Italy)
- M. Trimarchi (University of Messina, Italy)
-

Local Organizing Committee: A. Anastasi, L. Auditore, F. Curciarello, G. Mandaglio, L. Quattrocchi, A. Trifirò, M. Trimarchi.

Web-Site: http://newcleo.unime.it/Events/NRNN2017

List of speakers and participants

Antonio Cannuli (University of Messina, Messina, Italy)
Luigi Capponi (ELI-NP, Bucharest, Romania)
Fabio Crespi (INFN - Milano, Italy)
Enrico de Filippo (INFN - Catania, Italy)
Deqing Fang (SINAP, Shanghai, China)
Giorgio Giardina (University of Messina, Messina, Italy)
Brunilde Gnoffo (INFN - Catania, Italy)
Sophie Heinz (GSI, Darmstadt, Germany)
Gaetano Lanzalone (Un. Kore, Enna & INFN - LNS, Catania, Italy)
Giuseppe Mandaglio (University of Messina, Italy)
Nunzia Simona Martorana (INFN - LNS, Catania, Italy)
Paolo Napolitani (IPN, CNRS/IN2P3, Un. Paris Sud, Orsay, France)
Avazbek K. Nasirov (JINR, Dubna, Russia)
Francesco Noferini (INFN - Bologna, Italy)
Emanuele Pagano (INFN - LNS, Catania, Italy)
Concetta Parascandolo (INFN - Napoli, Italy)
Alessandro Pulcini (Univ. "Federico II", Naples, Italy)
Lucia Quattrocchi (INFN - Catania, Italy)
Daniele Quero (Univ. "Federico II", Naples, Italy)
Paolo Russotto (INFN - LNS, Catania, Italy)
Vazgen Sargsyan (JINR - BLTP, Dubna, Russia)
Adam Szczepaniak (Indiana Un., Bloomington, USA)
Marina Trimarchi (University of Messina, Italy)
Antonio Trifirò (University of Messina, Italy)

IOP Conf. Series: Journal of Physics: Conf. Series **1014** (2018) 011002 doi:10.1088/1742-6596/1014/1/011002

Emanuele Vardaci (Univ. "Federico II", Naples, Italy)
Fengshou Zhang (Beijing Normal University, Beijing Shi, China)
Angelo Pagano (INFN Sez. Catania, Italy)
Giuseppe Cardella (INFN Sez. Catania, Italy)

IOP Conf. Series: Journal of Physics: Conf. Series **1014** (2018) 011003 doi:10.1088/1742-6596/1014/1/011003

International Workshop

Nuclear Reactions on Nucleons and Nuclei

Messina (Italy) October 25-26, 2017

Organized by: University of Messina

Chairman: Giuseppe Mandaglio (Messina)
Chairman: Antonio Trifirò (Messina)
Chairman: Marina Trimarchi (Messina)

Honorary Chairman:
Giorgio Giardina (Messina)

Scientific Secretaries: Antonio Anastasi (Messina), Lucia Quattrocchi (Catania).

Topics:

Nuclear Reaction Dynamics
Heavy and Superheavy Elements Synthesis
Multifragmentation Reactions
Nuclear Reactions induced by Radioactive Beams
Exotic Nuclei and Nuclear Reactions of Astrophysical Interest
Properties of Nuclear Matter

International Advisory Committee:

G. Adamian (JINR, Dubna, Russia)
N.V. Antonenko (JINR, Dubna, Russia)
B.B. Back (ORNL, Oak Ridge, USA)
A. Badalà (INFN - Catania, Italy)
M. Battaglieri (INFN - Genova, Italy)
G. Cardella (INFN - Catania, Italy)
G. Giardina (Un. of Messina, Italy)
S. Heinz (GSI, Darmstadt, Germany)
E. Kozulin (JINR, Dubna, Russia)
Yu-Gang Ma (SINAP, Shangai, China)
G. Mandaglio (Un. of Messina, Italy)
Jie Meng (Peking Un., China)
D. Moricciani (INFN - Roma2, Italy)
A.K. Nasirov (JINR, Dubna, Russia)
K. Nishio (JAEA, Ibaraki, Japan)
Yu. Ts. Oganessian (JINR, Dubna, Russia)
A. Pagano (INFN - Catania, Italy)
P. Pedroni (INFN - Pavia, Italy)
K. Pomorski (UMCS, Lublin, Poland)
E. Prasad (Cent. Univ. Kerala, India)
F. Riggi (Un. of Catania, Italy)
F. Rizzo (Un. of Catania and LNS, Catania, Italy)
A. Sobiczewski (Un. of Warsaw, Poland)
A. Szczepaniak (Indiana Un., Bloomington, USA)
A. Trifirò (Un. of Messina, Italy)
M. Trimarchi (Un. of Messina, Italy)

Local Organizing Committee: A. Anastasi, L. Auditore, F. Curciarello, G. Mandaglio, L. Quattrocchi, A. Trifirò, M. Trimarchi.

Workshop Proceedings will be published in
Journal of Physics: Conference Series

Contact: loc@unime.it **Website**: http://newcleo.unime.it/Events/NRNN2017

IOP Conf. Series: Journal of Physics: Conf. Series **1014** (2018) 011004 doi:10.1088/1742-6596/1014/1/011004

Peer review statement

All papers published in this volume of *Journal of Physics: Conference Series* have been peer reviewed through processes administered by the proceedings Editors. Reviews were conducted by expert referees to the professional and scientific standards expected of a proceedings journal published by IOP Publishing.

Indoor Fast Neutron Generator for Biophysical and Electronic Applications

A Cannuli[1,2], M T Caccamo[1,5], N Marchese[3,4], E A Tomarchio[6], C Pace[3,4] and S Magazù[1]

[1] Dipartimento di Scienze Matematiche e Informatiche, Scienze Fisiche e Scienze della Terra, Università di Messina, V.le F. Stagno D'Alcontres 31, 98166 Messina, Italia

[2] INFN - LNS - Istituto Nazionale di Fisica Nucleare - Laboratori Nazionali del Sud - Via Santa Sofia 62, 95123 Catania, Italia

[3] Dipartimento di Ingegneria Informatica, Modellistica, Elettronica e Sistemistica, Università della Calabria, Via P. Bucci, 87036 Arcavacata di Rende, Cosenza, Italia

[4] INFN – LNF - Istituto Nazionale di Fisica Nucleare - Laboratori Nazionali di Frascati - Via Enrico Fermi 40, 00044 Frascati, Roma, Italia

[5] INDAM, Istituto Nazionale di Alta Matematica "F. Severi", P.le Aldo Moro 5, 00185 - Roma, Italia

[6] Dipartimento Energia, Ingegneria dell'Informazione e Modelli Matematici, Università di Palermo, V.le delle Scienze, Ed. 9, 90128 Palermo, Italia

acannuli@unime.it

Abstract. This study focuses the attention on an indoor fast neutron generator for biophysical and electronic applications. More specifically, the findings obtained by several simulations with the MCNP Monte Carlo code, necessary for the realization of a shield for indoor measurements, are presented. Furthermore, an evaluation of the neutron spectrum modification caused by the shielding is reported. Fast neutron generators are a valid and interesting available source of neutrons, increasingly employed in a wide range of research fields, such as science and engineering. The employed portable pulsed neutron source is a MP320 Thermo Scientific neutron generator, able to generate 2.5 MeV neutrons with a neutron yield of 2.0×10^6 n/s, a pulse rate of 250 Hz to 20 KHz and a duty factor varying from 5% to 100%. The neutron generator, based on Deuterium-Deuterium nuclear fusion reactions, is employed in conjunction with a solid-state photon detector, made of n-type high-purity germanium (PINS-GMX by ORTEC) and it is mainly addressed to biophysical and electronic studies. The present study showed a proposal for the realization of a shield necessary for indoor applications for MP320 neutron generator, with a particular analysis of the transport of neutrons simulated with Monte Carlo code and described the two main lines of research in which the source will be used.

1. Introduction

For industrial applications that require neutrons, it is possible the employ of three different primary sources, such as nuclear reactors, radioisotopes and accelerator-based neutron sources. Nuclear reactors are clearly the largest and most prolific sources of neutrons, though their size, complexity and cost have limited their industrial applications for purposes other than electricity generation. In contrast, radioisotope neutron sources are employed in a myriad of industrial applications, including thickness gauging and petroleum exploration. Furthermore, radioisotope sources are ideal for fixed installations that run continuously, but they are not well suited for applications that require pulsed neutrons and they may create safety issues and logistical complications as well. Accelerator-based neutron sources are the third available source of neutrons for industrial applications. These systems vary in size and diversity and they include large installations, for example the spallation neutron source and the smaller photo-neutron sources. Among these accelerators, compact devices, known as fast Neutron Generators (NG)s are mostly employed in different fields that require neutrons. They are designed as hermetic, sealed tubes and use the following Deuterium–Deuterium (D–D) or Deuterium–Tritium (D–T) reactions:

$$D + D \rightarrow {}^{3}_{2}He + {}^{1}_{0}n + 3.270MeV \tag{1}$$

$$D + T \rightarrow {}^{4}_{2}He + {}^{1}_{0}n + 17.590MeV. \tag{2}$$

They are able to generate neutrons of an energy of 2.5 MeV in the case of D-D reactions and 14.1 MeV in the case of D-T reactions. In general, a fast NG is constituted by a source for generating positively charged ions; one or more devices to accelerate these ions; a metal hydride target loaded with either deuterium, tritium, or a mixture of the two and a gas-control reservoir, also made of the same material.

In this study the findings obtained by several simulations with MCNP Monte Carlo code, necessary for the realization of a shield for an indoor fast NG for biophysical and electronic applications, are presented. Furthermore, an evaluation of the neutron spectrum modification caused by the shielding is reported. The obtained results showed a proposal for the realization of a shield necessary for indoor applications for a MP320 NG, with a particular analysis of the transport of neutrons simulated with Monte Carlo code and described the two main lines of research in which the source will be used [1-10].

2. Fast Neutron Generator and shielding design calculations

Fast NGs are valid and interesting available sources of neutrons, increasingly employed in a wide range of research fields, such as science and engineering. The Thermo Scientific MP 320 is an example of very lightweight portable fast NG. It can be used in different applications like explosive, weapon detection, or drug detection and it can be employed in laboratories for different kind of research applications because it is a relatively compact system that requires a very low power. It is possible to operate with either a D-T or a D-D neutron tube, simply changing the target inside the neutron tube. In fact, in "D-T mode", the source is able to generate a neutron yield of 1.0×10^{8} n/s with an energy of 14.1 MeV, while in "D-D mode" it generates a neutron yield of 2.0×10^{6} n/s with an energy of 2.5 MeV. The NG can be modulated varying the pulse rate from 250 Hz to 20 kHz, varying respectively from 5% to 100% the duty factor. It can be also used in conjunction with an ORTEC PINS-GMX solid-state photon detector, made of n-type high-purity germanium. For indoor application, such as research activities, the NG needs the realization of a specific bunker necessary to avoid damage from radiation exposition for users and instrumentation. Neutrons interact with nuclei in a variety of ways, for instance, if the nucleus is unchanged in either isotopic composition or internal energy after interacting with a neutron, the process is called elastic scattering. On the other hand, if the nucleus, still unchanged in composition, is left in an excited state, the process is called inelastic scattering. Generally, two or more neutrons are emitted when a nucleus is struck by a high energy neutron. There are three different procedures for the calculation of the shielding, which are neutrons transport equation, the Monte Carlo simulation and the removal cross section method.

Neutrons transport theory is relatively simple in principle and it is possible to determine an exact equation for transport phenomena, i.e. Boltzmann equation, also known as transport equation. Unfortunately, it is much easier to derive the Boltzmann equation than it is to solve it, though under some specific conditions, such as uniform materials and same energy for the neutrons, the equation simplifies considerably the problem and it can be calculated in a rather straightforward way. This simplified transport theory is known as neutron diffusion equation and it is shown in the following:

$$D\,\nabla^2\phi - \Sigma_a\phi + s = \frac{1}{v}\frac{\partial\phi}{\partial t}. \quad (3)$$

However, many diffusion problems can not be solved by the analytical techniques; they have need of numerical computation methods [11].

Another different method shielding calculation is the Monte Carlo simulation. It consists of simulating individual particles and recording some aspects of their average behavior, using the central limit theorem. Monte Carlo simulation can be employed to duplicate theoretically a statistical process for complex problems that can not be modeled by computer codes, using deterministic methods. The individual probabilistic events of a process are simulated sequentially and the probability distributions, which govern the events, are statistically sampled to describe the total phenomenon.

Finally, the removal cross section method takes into account the ability of materials to remove neutrons of different energies from the primary beam. The interactions that slow neutrons down and cause their eventual removal from a beam are probabilistic, whereby a flux of neutrons of intensity I will be diminished in a thickness x of absorber proportionally to the intensity of the neutron source and the neutron removal coefficient Σ_{nr}, of the absorbing material:

$$-\frac{dI}{dx} = \Sigma_{nr}I. \quad (4)$$

The solution of the equation is:

$$I(x) = I_0e^{-\Sigma_{nr}x} \quad (5)$$

in which, I_0 is the initial intensity, $I(x)$ refers to those neutrons that penetrate a distance x in the absorber without a collision, $e^{-\Sigma_{nr}x}$ is the probability that a given neutron travels at a distance x without any interaction. From a conceptual point of view, Σ_{nr} can be considered as the probability per unit path length that a neutron will undergo an interaction as it moves through an absorber and it will be removed from the beam by either absorption or scattering [11].

3. Experimental results for Thermo Scientific MP320 NG shielding

For the design of the MP320 NG shielding, the features in table 1 were taken as a reference.

Table 1. Features for the design of the MP320 NG shielding.

reaction	D-D
max neutron energy [MeV]	2.45
neutron yield [n s^{-1}]	2.00E+06

In order to perform a preliminarily evaluation of the shielding thicknesses was used the method of removal cross section. In a second moment, data preliminarily obtained were verified with MCNP Monte Carlo code simulations. The removal cross sections for different materials are available in several handbooks for shielding dimensioning, in particular, the employed materials for neutron shielding and their Σ_r are reported in table 2 [12-14].

Table 2. Materials employed for neutron shielding.

material	**ρ [g cm^{-3}]**	**Σ_r [cm^{-1}]**
borated silicone (Elmahroug et al, 2013)	1.59	0.1007
30% borated PE (Elmahroug et al, 2013)	1.19	0.1191
iron (Kaye and Laby, 2005)	7.86	0.168
lead (Kaye and Laby, 2005)	11.35	0.116
barite concrete (NBS, 1957)	3.491	0.0945

In order to optimize the shielding size it is necessary to select materials that offer a good shielding power, i.e. an high Σ_r. It is also important to utilize the interactions of neutrons with the different materials in order to optimize the processes of slowing down and capture. Therefore the shielding is made of successive layers of different materials. The exponential attenuation law is valid only for collimated neutron beam, while in the case in which the source can be considered as a point source, the spatial dependence of the neutron flux has to be considered. Therefore the attenuation law becomes the following:

$$\Phi = \frac{\varphi_0 e^{-\Sigma_r \cdot t}}{4\pi d^2} \tag{6}$$

where φ_0 is the initial neutron yield (n s^{-1}), d is the distance at which neutron flux is calculated respect to the source position, Φ is the neutron flux at distance d from the source and t is the thickness of shielding material. The method of removal cross section allows to obtain realistic estimations of the neutrons attenuation. Further evaluations can be performed using numerical codes based on the Monte Carlo method. The aim of the shielding is to reduce the neutron flux up to make the dose rate lower than the project limit. The evaluation of the different materials thickness was made considering their characteristics (Σ_r and weight). Therefore, the materials and thickness in table 3 were chosen.

Table 3. Materials and thickness for Neutron Generator MP320 shielding.

material	**Σ_r [cm^{-1}]**	**t [cm]**
30% borated PE	0.1191	15
barite concrete	0.0945	15

IOP Conf. Series: Journal of Physics: Conf. Series **1014** (2018) 012001 doi:10.1088/1742-6596/1014/1/012001

Neutron flux was calculated on the external surfaces at the end of the shielding, taking in consideration the distance between the source, considered a point source, the internal surface of the shielding and the entire shielding thickness. The values of neutron flux obtained for the different surfaces are shown in table 4.

Table 4. Neutron flux values obtained for the different surfaces of the shielding.

	shielding							
	up		lateral		back		front	
material	t [cm]	$\acute{\Phi}$ [n cm^{-2} s^{-1}]	t [cm]	$\acute{\Phi}$ [n cm^{-2} s^{-1}]	t [cm]	$\acute{\Phi}$ [n cm^{-2} s^{-1}]	t [cm]	$\acute{\Phi}$ [n cm^{-2} s^{-1}]
air	33	146.15	15	397.89	23	300.86	57	48.99
30% borated PE	15	11.57	15	21.77	15	18.47	30	0.59
barite concrete	15	1.63	15	2.59	15	2.30	0	0.59

In figure 1, it is possible to see the different thicknesses of the shielding: the green part represents the 30% borated PE, while the grey part, barite concrete.

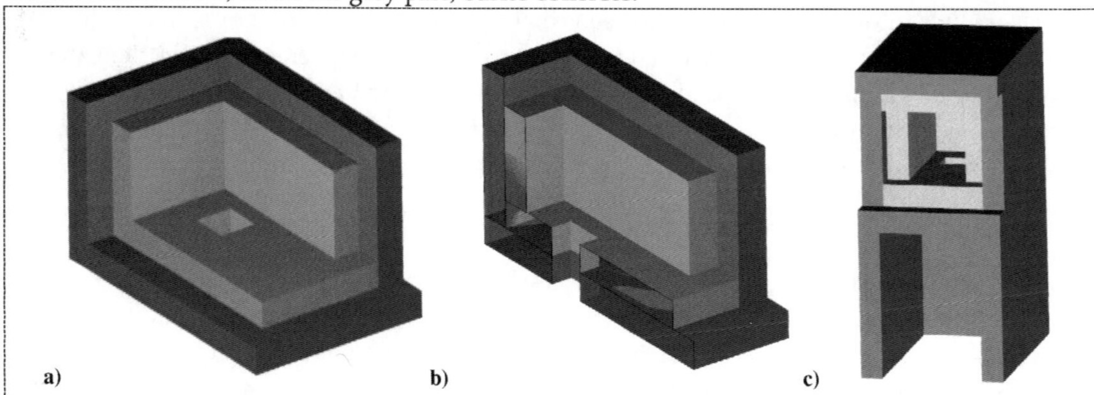

Figure 1. Representation of shielding thicknesses (a), shielding section view (b) and shielding structure (c).

In the section view, it is possible to identify the window for the samples irradiation. NG will be located inside the shielding, with the neutron emitting target in correspondence of the window. The entire shielding will be sustained by a concrete structure as shows in figure 1 (c). Figure 2 shows the shielding inside the laboratory and the comparison between the shielding and a user.

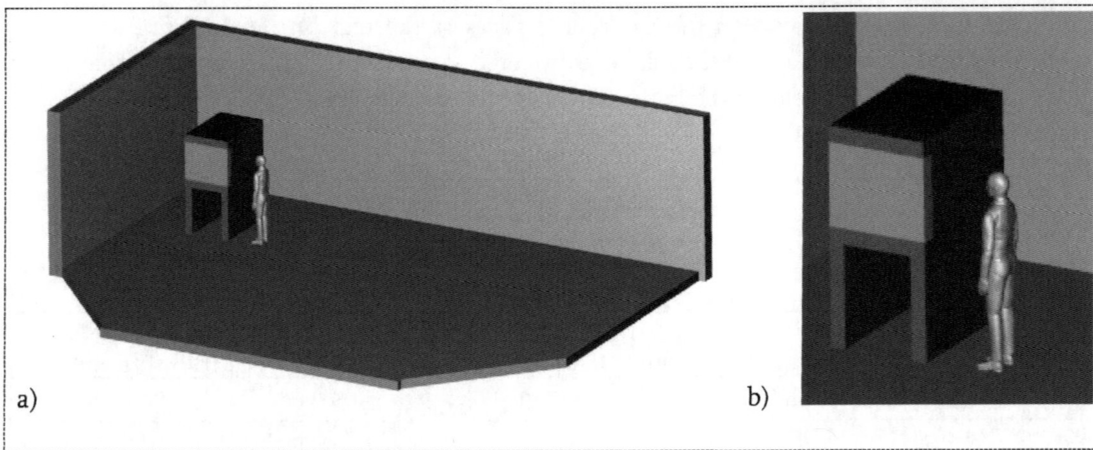

Figure 2. Shielding representation inside the laboratory (a) and comparison between the shielding and a user.

Several simulations were performed with a software based on Monte Carlo method, in order to verify the values obtained with the application of removal cross section. The code used was MCNP5. The code include information relatively to the geometry, materials and source characteristics. The cells in which evaluate the neutron flux were also specified. These were placed on the external surfaces of the shielding in correspondence of the specification addressed in the design phase. For neutron flux evaluation was chosen the Track Length Estimate of Cell Flux (F4) that allow to obtain the average neutron flux in a volume V, as described in following formulation:

$$\overline{\phi_V} = \frac{1}{V}\int dE \int dV \int ds\, N(\vec{r}, E, t) \quad (7)$$

where $N(\vec{r}, E, t)$ is the density of particles, regardless of their trajectories, at a point.

Thus, the code studies the life of the particles generated into the source and then evaluates their average behavior in the volume V. Therefore, taking into consideration the complexity of the problem, it will be necessary to simulate an adequate number of particles in order to reduce the uncertainty on the flux values.

Two sections of the shielding are reported in figure 3: the blue part represents 30% Borate PE, the yellow part, barite concrete and the green part, the cells in which neutron flux was calculated. The cross in the two figures shows the point at which the "neutron source" is located.

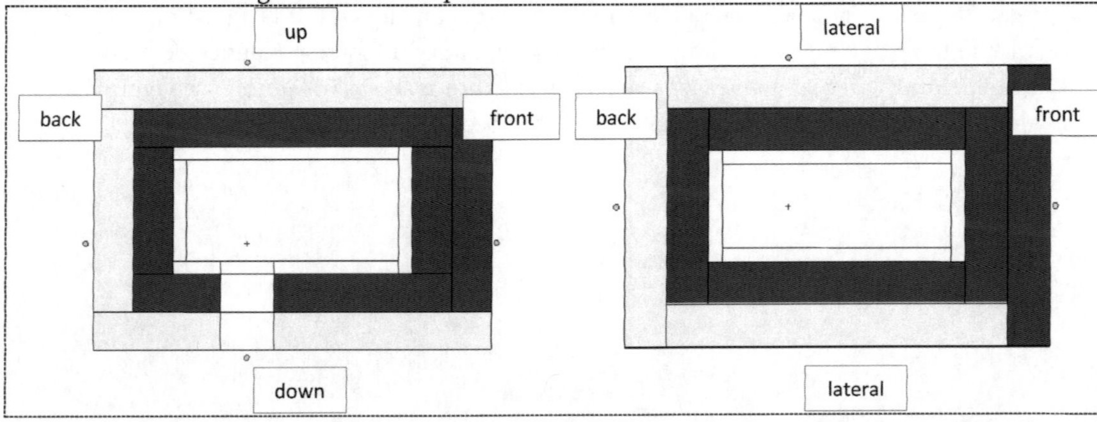

Figure 3. Shielding sections from MCNP simulations.

Simulations were performed considering the life of $2 \cdot 10^9$ neutrons, in order to evaluate the flux in the green cells and in order to obtain a standard deviation under the 10%. The values obtained by the simulation represent the average flux into the cells volume for every neutron started from the source.

They must be handled in order to obtain the value of the mean flow in the cells. After handling the values, the results in table 5 were obtained.

Table 5. Neutron flux values obtained by Monte Carlo simulation.

	shielding			
	up	**lateral**	**back**	**front**
energy [MeV]	Φ **[n cm^{-2} s^{-1}]**	Φ **[n cm^{-2} s^{-1}]**	Φ **[n cm^{-2} s^{-1}]**	Φ **[n cm^{-2} s^{-1}]**
2.45	0.349	0.103	0.252	8.87×10^{-3}
total	3.75	1.61	2.63	0.145

Comparing the values obtained from the simulation with those obtained by the removal cross section method, as shown in table 6, it is possible to verify that this method allows to obtain the realistic values of the neutron flux and it represents a valid approach to determine neutron shielding thicknesses.

Table 6. Comparison between Removal Cross Section Method and Monte Carlo Method.

	shielding			
	up	**lateral**	**back**	**front**
	Φ **[n cm^{-2} s^{-1}]**	Φ **[n cm^{-2} s^{-1}]**	Φ **[n cm^{-2} s^{-1}]**	Φ **[n cm^{-2} s^{-1}]**
Σ_r **method**	1.63	2.59	2.30	0.59
MCNP total	3.75	1.61	2.63	0.145

In order to evaluate the effect of shielding, it is possible to calculate the neutron dose rate. The following parameters were taken into account as references for the dose rate limits evaluation:

• Project Equivalent Dose [mSv/y] 0.5;
• Annual duty hours [h/y] 1.200.

The project neutron dose rate was chosen as half of the annual limit for the population in order to consider also the photons contribute to the total dose rate. The annual duty hours were chosen as the typical lifetime of the target. Taking into account of the set parameters, the reference value for the dose rate is equal to:

$$\frac{0.5\left[\frac{mSv}{y}\right]}{1{,}200\left[\frac{h}{y}\right]} = 4.2 \cdot 10^{-4}\left[\frac{mSv}{h}\right]. \tag{8}$$

In order to evaluate the neutron dose rate, it is necessary to know the neutron spectrum and to apply the flow-to-dose rate energy-dependent conversion factors. Considering the neutron spectrum obtained with MCNP simulations and applying the conversion factors, the dose rate values, shown in table 7, were obtained [15-16].

Table 7. Equivalent Dose Rate values outside the neutron shielding.

	shielding			
	up	**lateral**	**back**	**front**
equivalent dose rate [mSv/h]	$1.51\text{x}10^{-3}$	$2.81\ \text{x}10^{-2}$	$4.0\text{E}\ \text{x}10^{-2}$	$1.62\ \text{x}10^{-2}$
equivalent dose rate @ 0.1 m from the shielding [mSv/h]	$1.20\ \text{x}10^{-8}$	$2.23\ \text{x}10^{-7}$	$3.18\ \text{x}10^{-7}$	$1.29\ \text{x}10^{-7}$

The obtained values of neutron dose rate are over the limit of $4.2 \cdot 10^{-4}[mSv/h]$, however, taking into account the geometric attenuation is possible to evaluate that the neutron dose rate goes down rapidly below the limit.

Every application field of neutron sources needs particular characteristics of irradiation and it is necessary to evaluate if the source is suitable for the tests. The shielding or other structural materials cause the modification of the energy spectrum of a neutron source and, hence, of the exposed samples. Therefore it is necessary to evaluate the modified energy spectrum in order to define the real characteristics of the neutron source. Several simulations were performed to this aim, using MCNP. The neutron spectrum was evaluated in correspondence to the window for the irradiation of the samples, outside the shielding, as shown in figure 4.

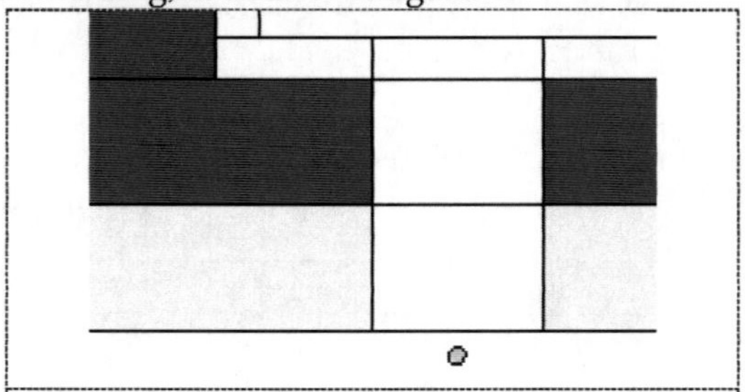

Figure 4. Detail of spectrum evaluation point.

The simulations were performed both for the D-D and D-T mode in order to evaluate the effect of shielding on the two different energy neutron beams.

IOP Conf. Series: Journal of Physics: Conf. Series **1014** (2018) 012001 doi:10.1088/1742-6596/1014/1/012001

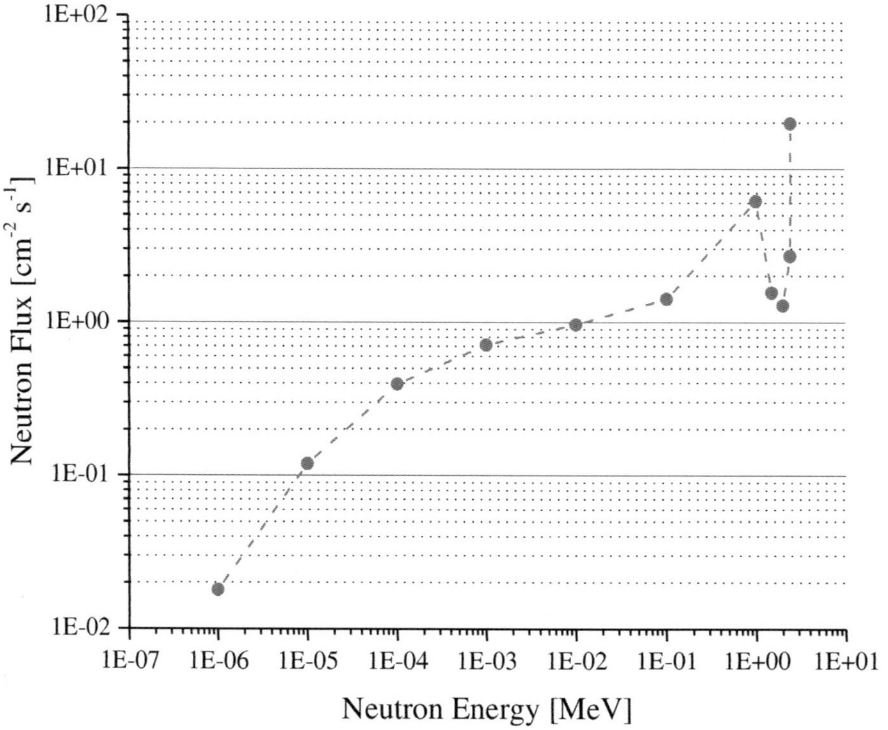

a)

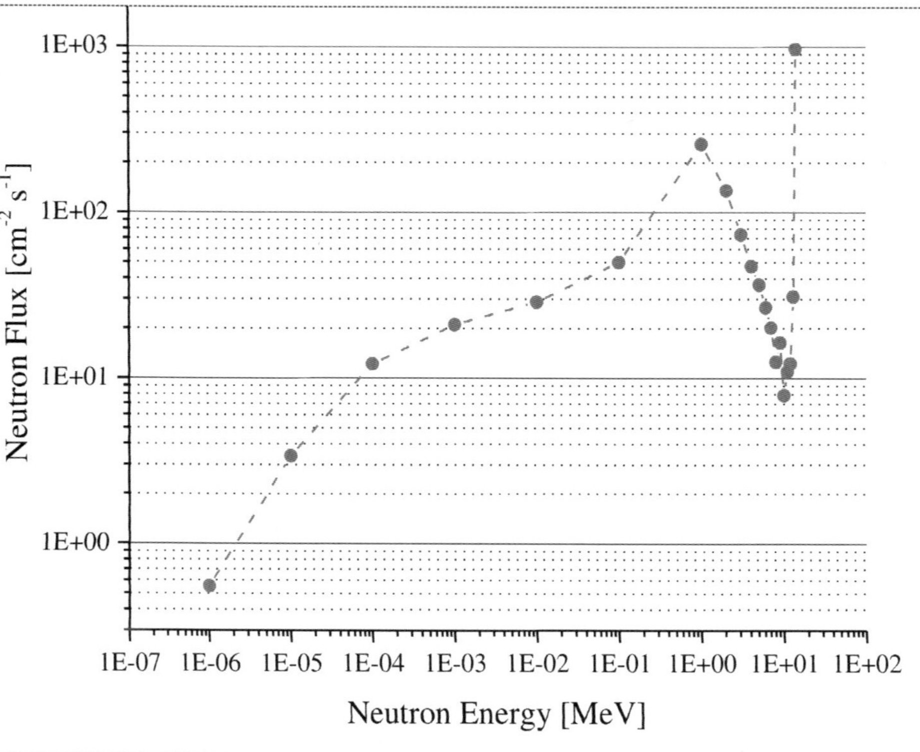

b)

Figure 5. a) D-D neutron spectrum; b) D-T neutron spectrum.

IOP Conf. Series: Journal of Physics: Conf. Series **1014** (2018) 012001 doi:10.1088/1742-6596/1014/1/012001

The graphics in figure 5 show how the presence of the shielding contributes to the spectrum change to which the samples will be exposed.

4. Biophysical and electronic applications

The indoor MP320 NG can be employed both for biophysical and electronic applications, some example are elemental analysis and testing of devices. Elemental analysis, known as Neutron Activation Analysis (NAA) is an analytical technique for qualitative and quantitative multi-elemental analysis of major, minor and trace elements. For example, with this technique, it is possible to trace heavy metals in samples of biophysical interest that contain aqueous solutions [17-21], disaccharides [22-39] or polymers [40-47] and which have already been studied with other spectroscopic techniques [48-51]. In NAA, a target, after a non-elastic collision in the atomic nucleus, absorbs a neutron and forms a compound in an excited state. When the compound nucleus de-excites into a more stable configuration, an emission of one or more characteristic prompt gamma rays occurs. The process ends when the ground state formed during the de-excitation, is stable, i.e. at the moment in which an isotope of the same element, with a mass number increased by one, is produced. Generally, every chemical element has at least one isotope that produces a radioactive one in neutron capture. They typically emit β particles and γ radiation with a given half-life. The energies of the prompt and delayed gamma rays allow to identify the emitter nuclide and their intensity is proportional to its amount.

The Earth's atmosphere acts as a shield against ionizing radiation coming from deep space. However, the interaction of galactic cosmic rays with the terrestrial atmosphere produces a "shower" of secondary particles, mainly neutrons, which come down to sea level [52]. Thus, all electronic devices are exposed to this "natural" radiation environment. Despite the neutron flux is quite low (especially at sea level) even a single neutron, under certain conditions, can cause a failure into electronic devices: this case is known as Single Event Effects (SEE)s [53]. SEEs are related to the charge generated in the semiconductor, through ionization realized by the incident particle. In the case of neutrons, this mechanism is activated by the emission of secondary charged particles, after a nuclear reaction, radioactive decay or spallation [54]. If the ionization takes place near a reverse biased pn junction, the electron-hole pairs are separated by the electric field and then collected by the electrodes, generating a transient current. This microscopic mechanism can lead to different macroscopic effects, as for example, Single Event Upset (SEU) in a memory [55] or Single Event Burnout (SEB) affecting power MOSFETs [56]. In order to characterize a device's sensitivity to terrestrial neutrons, it is necessary to perform accelerated tests with dedicated neutron facilities. The procedures to carry out such tests and the required featured of the facilities are regulated by international standards such as the JESD89A. The scope of these tests is to assess the failure rate associated with terrestrial neutron SEE.

Spallation and mono-energetic sources are suggested for testing devices with high-energy neutrons. Spallation sources produce a "white" neutron spectrum, which mimics the natural terrestrial neutron environment. A valid alternative could be represented by mono-energetic facilities. There are only few mono-energetic neutron sources: D-T and D-D facilities, respectively at 14 MeV and 2.45 MeV. According with the standard JESD89A. In this context, the proposed neutron facility could be used for accelerated SEE testing of electronic devices.

Conclusions

IOP Conf. Series: Journal of Physics: Conf. Series **1014** (2018) 012001 doi:10.1088/1742-6596/1014/1/012001

The data obtained with the method of removal cross section and through simulations with MCNP represent a fundamental aid in the design of the shield, necessary for the indoor use of the neutron source. This work is therefore a starting point for the construction of the shield and the obtained data will be verified experimentally. The project could be subjected to modifications in order to optimize the irradiation characteristics of the samples. The employed predictive tools will be so critical to the optimization process and will allow to intervene on the shielding before its construction. The whole device represent a valid research tool and the areas where it can be used are numerous and biophysics and electronics represent only two examples of the many possible uses of this facility. The reduced operating costs, the compactness and the possibility of varying the duty factor make the neutron generator a very competitive device within the panorama of the available neutron sources.

References

[1] Csikai J 1987 *CRC Handbook of Fast Neutron Generators* (Boca Raton: CRC Press)
[2] James W D 2000 *J. Radioanal. Nucl. Chem.* **243** 119
[3] Marchese N, Cannuli A, Caccamo M T and Pace C 2016 *Biochim. Biophys. Acta* 1861 3661-3670
[4] Chichester D L and Simpson J D 2003-2004 *The Industrial Physicist, American Institute of Physics* **9** 22
[5] Reijonen J 2005 Compact Neutron Generators for Medical, Home Land Security and Planetary Exploration, *Proceedings of the IEEE* 49-53
[6] Ludewigt B A, Wells R P and Reijonen J 2007 *Nucl. Instrum. Methods Phys. Res. B* **261** 830-834
[7] Das B K and Shyam A 2008 *Rev. Sci. Instrum.* **79** 123305
[8] Hennet L *et al.* 2011 *Eur. Phys. J. Spec. Top.* 196 151-165
[9] Liang L, Rinaldi R and Schober H 2009 *Neutron Applications in Earth, Energy and Environmental Sciences* (New York: Springer)
[10] Liu W, Li M, Gao K and Gu D 2014 *Nucl. Instrum. Methods Phys. Res. A* **768** 120–123
[11] Lamarsh J R 1966 Introduction to Nuclear Reactor Theory. *Addison-Wesley Publishing Company* 17-118-132-152
[12] National Bureau of Standards 1957 *Protection against neutron radiation up to 30 Million electron Volts* **63** 81
[13] Kaye G W C and Laby T H 1995 *Tables of Physical & Chemical Constants Kaye & Laby Online*
[14] Elmahroug Y, Tellili B and Souga C. 2013 *Intern. J. Phys. Res.* **3** 11
[15] Seog-Guen K *et al.* 1980 *J. Korean Phys. Soc.* 12 173
[16] Ward D C 2009 Impact of switching to the ICRP-74 neutron flux-to-dose equivalent rate conversion factors at the sandia national laboratory building 818 neutron source range 1144
[17] Lokotosh T V, Magazù S, Maisano G and Malomuzh N P 2000 *Phy.s Rev. E – Stat. Physics, Plasmas, Fluids, Relat. Interdiscip.* **62** 3572-3580
[18] Magazù S 1996 *Phys. B.* **226** 92-106
[19] Magazù S, Maisano G, Mallamace F and Micali N 1989 *Phys. Rev. A* **39** 4195-4200
[20] Jannelli M P, Magazù S, Migliardo P, Aliotta F and Tettamanti E 1996 *J. Phys. Condens. Matter.* **8** 8157-8171
[21] Magazù S, Migliardo F and Telling M T F 2008 *Food Chem.* **106** 1460-1466
[22] Magazù S, Calabrò E, Caccamo M T and Cannuli A 2016 *Curr. Chem. Biol.* **10** 57-64
[23] Magazù S, Maisano G, Migliardo P, Middendorf H D and Villari V 1998 *J Chem Phys.* **109** 1170-1174
[24] Magazu S *et al.* 1998 *J. Phys. Chem. B* **102** 2060-2063
[25] Pagnotta S E, Ricci M A, Bruni F, McLain S and Magazù S 2008 *Chem. Phys.* **345** 159-163
[26] Magazù S, Migliardo F and Telling M T F 2007 *Eur Biophys J.* **36** 163-171
[27] Minutoli L *et al.* 2007 *Shock* **27** 91-96
[28] Magazù S, Calabrò E and Campo S 2010 *J. Phys. Chem. B* **114** 12144-12149

[29] Barreca D *et al.* 2010 *Biophys. Chem.* **147** 146-152
[30] Migliardo F, Caccamo MT and Magazù S 2014 *Food Biophys.* **9**
[31] Magazù S, Migliardo F, Vertessy B G and Caccamo M T. 2013 *Chem Phys.* 424
[32] Migliardo F, Caccamo M T and Magazù S 2013 *J. Non Cryst. Solids.* 378
[33] Magazù S, Migliardo F, Vertessy B G and Caccamo M T 2013 *Chem. Phys.* **424** 56-61
[34] Branca C, Magazù S, Maisano G, Bennington A and Fåk B 2003 *J. Phys. Chem. B* 107 1444-1451
[35] Barreca D, Lagana G, Bruno G, Magazù S and Bellocco E 2013 *Biochim.* **95** 2042-2049
[36] Branca C, Magazù S, Migliardo F and Migliardo P 2002 *Phys. A* **304** 314-318
[37] Branca C *et al.* 1999 *J. Phys Cond. Matter.* **11** 3823-3832
[38] Magazù S *et al.* 2013 Nova Science Pub. 35-46
[39] Branca C *et al.* 1999 *J. Phys. Cond. Mat.* **11** 3823-3832
[40] Caccamo M T, Cannuli A, Calabrò E and Magazù S 2017 *IOP Conf. Ser. Mater. Sci. Eng.* **199** 12119
[41] Faraone A, Magazù S, Maisano G, Ponterio R and Villari V 1999 *Macromol.* **32** 1128-1133
[42] Branca C, Magazù S, Maisano G, Migliardo P and Villari V 1998 *J. Phys. Condens. Matter.* **10** 10141-10157
[43] Migliardo F, Magazù S and Caccamo M T 2013 *J. Mol. Struct.* **1048** 261-266
[44] Caccamo M T and Magazù S 2016 *Vib. Spectrosc.* **85** 222-227
[45] Caccamo M T and Magazù S 2017 *Appl. Spectrosc.* **71** 401-409
[46] Caccamo M T and Magazù S 2017 *Spectrosc. Lett.* **50** 130-136
[47] Caccamo M T and Magazù S 2017 *Polym. Testing* **62** 311-318
[48] Magazù S, Migliardo F and Caccamo M T 2013 *Adv. Mater. Sci. Eng.* 1-7
[49] Caccamo M T and Magazù S 2017 *Eur. J. Phys.* **38**
[50] Caccamo M T, Calabrò E, Cannuli A and Magazù S 2016 *MATEC Web Conf.* **55** 02004
[51] Magazù S, Migliardo F and Caccamo M T 2012 *J. Phys. Chem. B* **116** 9417-9423
[52] Ziegler J F 1998 *IBM J. Res. Develop.* **42** 117-140
[53] Leray J L 2007 *Microelec. Rel.* **47** 1827-1835
[54] Tang H H K 1996 *IBM J. Res. Develop.* **40** 91-108
[55] Hands A, Morris P, Dyer C, Ryden K, and Truscott P 2011 *IEEE Trans. Nucl. Sci.* **58** 2687-2694
[56] Oberg D L, Wert J L, Normand E and Wender S A 1996 *IEEE Trans. Nucl. Sci.* **43** 2913-2920

IOP Conf. Series: Journal of Physics: Conf. Series **1014** (2018) 012002 doi:10.1088/1742-6596/1014/1/012002

Gamma decay of pygmy states in ^{90,94}Zr from inelastic scattering of light ions

F.C.L. Crespi[1], A. Bracco[1], A. Tamii[2], N. Blasi[3], F. Camera[1], O. Wieland[3], N. Aoi[2], D. Balabanski[4], S. Bassauer[5], A. S. Brown[6], M. P. Carpenter[7], J. J. Carroll[8], M. Ciemala[9], A. Czeszumska[2], P. J. Davies[6], L. Donaldson[10], Y. Fang[2], H. Fujita[2], G. Gey[2], T. H. Hoang[2], N. Ichige[11], E. Ideguchi[2], A. Inoue[2], J. Isaak[2], C. Iwamoto[12], D. G. Jenkins[6], O. H. Jin[2], T. Klaus[5], N. Kobayashi[2], T. Koike[11], M. Krzysiek[4,9], M. Kumar Raju[2], M. Liu[13], A. Maj[9], D. Montanari[14], L. Morris[6], S. Noji[15], S. G. Pickstone[16], D. Savran[17], M. Spieker[16,*], G. Steinhilber[5], C. Sullivan[18,19,20], B. Wasilewska[9], V. Werner[5], T. Yamamoto[2], Y. Yamamoto[2], X. Zhou[13], S. Zhu[7]

[1] Università degli Studi di Milano and INFN, Milano, I-20133, Italy
[2] Osaka Univ., RCNP, Ibaraki, Osaka 5670047, Japan
[3] INFN Sez. di Milano, Milano, I-20133, Italy
[4] Extreme Light Infrastructure - Nuclear Physics, "Horia Hulubei" National Institute for R&D in Physics and Nuclear Engineering, 30 Reactorului Str., 077125 Bucharest-Magurele, RomaniaExtreme Light Infrastruct. Nucl. Phys., 30 Reactorului, Bucharest 077125, Romania
[5] Tech. Univ. Darmstadt, Inst. Kernphys., D-64289 Darmstadt, Germany
[6] Univ. of York, Dept. Phys., York YO10 5DD, N Yorkshire, England
[7] Argonne Natl. Lab., Div. Phys, Argonne, IL 60439 USA
[8] US Army Res. Lab., Adelphi, MD 20783 USA
[9] Institute of Nuclear Physics Polish Academy of Sciences, ul. Radzikowskiego 152, 31-342 Krakow, Poland

[10] Univ. Witwatersrand, Sch. Phys, ZA-2050 Johannesburg, South Africa
[11] Tohoku Univ., Dept. Phys., Sendai, Miyagi 9808578, Japan
[12] Center for Nuclear Study (CNS), University of Tokyo
[13] Institute of Modern Physics, Chinese Academy of Sciences, Lanzhou, China
[14] Univ. Strasbourg, CNRS IN2P3, Inst. Plur. Hubert Curien, Strasbourg, France
[15] FRIB, Michigan State University, South Shaw Lane, East Lansing, MI 48824
[16] Univ. of Cologne, Inst. Nucl. Phys., Zulpicher Str 77, D-50937 Cologne, Germany
[17] GSI Helmholtzzentrum Schwerionenforsch. GmbH, EMMI, D-64291 Darmstadt, Germany
[18] Michigan State Univ, Natl. Supercond. Cyclotron Lab., E Lansing, MI 48824 USA
[19] Michigan State Univ, Joint Inst. Nucl. Astrophys., Ctr. Evolut. Elements, E. Lansing, MI 48824 USA
[20] Michigan State Univ., Dept. Phys. & Astron., E Lansing, MI 48824 USA

IOP Conf. Series: Journal of Physics: Conf. Series **1014** (2018) 012002 doi:10.1088/1742-6596/1014/1/012002

**Present address: Michigan State Univ, Natl. Supercond. Cyclotron Lab., E Lansing, MI 48824 USA*

fabio.crespi@mi.infn.it

Abstract. We performed experiments to study the low-energy part of the E1 response (Pygmy Dipole Resonance) in ^{90,94}Zr nuclei, by measuring the $(p,p'\gamma)$ and $(\alpha,\alpha'\gamma)$ inelastic scattering reactions at energies $E_{beam,p}$ = 80 MeV and $E_{beam,\alpha}$ =130 MeV respectively. The inelastically scattered particles were measured by employing the high-resolution spectrometer Grand Raiden. The gamma-rays emitted following the de-excitation of the Zr target nuclei were detected using both the clover type HPGe detectors of the CAGRA array and the large volume $LaBr_3$:Ce scintillation detectors from the HECTOR+ array. Some preliminary results are presented here.

1. Introduction

It is well known that the electric dipole strength in nuclei is dominated by the Giant Dipole Resonance. This collective mode is perceived as an oscillation of all the neutrons against all the protons. However, it is found the systematic presence of an accumulation of electric dipole strength, at energies around the particle threshold, in most of the studied neutron rich nuclei. To this accumulation of strength has been given the name of Pygmy Dipole Resonance [1,2]. The word pygmy was used due to the much smaller fraction of energy-weighted sum rule exhausted by this group of E1 states as compared to that exhausted by the GDR. The PDR is possibly associated to a new kind of collective mode arising from the oscillation of the N=Z core against the neutron skin in neutron rich nuclei. The properties of the PDR strength are connected to the properties of the neutron skin which in turn are used to constrain the equation of state of neutron rich matter [3,4,5]. In addition, the PDR strength plays a role for r-process nucleosynthesis [6].

From the experimental point of view an important finding concerns the observation of the structural splitting of the PDR states into two separated groups [7,8]. One higher-lying group of states has isovector nature and it is excited only in photon scattering experiments, these states are associated to the transition toward the GDR. One lower lying group of states of more isoscalar character is excited both in photon scattering and alpha scattering experiments. In addition experiments performed for the nuclei ^{208}Pb, ^{124}Sn, ^{90}Zr and ^{140}Ce [9-12] have shown that the $(^{17}O,^{17}O'\gamma)$ reaction is a good tool to investigate the isospin properties of the pygmy states.

Although a large number of experimental and theoretical works have been devoted to this subject in the past decades, the nature of the PDR is not well understood yet. Key issues for the interpretation of the nature of the PDR states are the determination of their characteristic transition density and the isospin character.

We proposed the experiments described here to study the low-energy part of the E1 response, by measuring the $(p,p'\gamma)$ and $(\alpha,\alpha'\gamma)$ inelastic scattering reactions with the CAGRA-Grand Raiden set-up. Our aim is to investigate the isospin character of the PDR states by comparing with high resolution data their population with the $(\alpha,\alpha'\gamma)$ reaction (isoscalar character, surface-sensitive), $(p,p'\gamma)$ reaction (mainly isoscalar character, better sensitivity to the inner transition density), and (γ,γ') (mainly isovector character).

The ^{90}Zr and ^{94}Zr nuclei were chosen because of their pronounced proton subshell closure and because they are rather well studied with the (γ, γ') reaction [13], [14]. The works with (γ,γ') reported in [13], [14] show clearly an excess of E1 yield at low energy as compared with the expectation using the Lorentzian tail of the GDR.

2. Experimental Technique and Setup

The experiments have been performed at the Research Center for Nuclear Physics (RCNP) of the Osaka University. The used beams of alpha particles and protons, at bombarding energy of respectively 130 MeV and 80 MeV, were provided by the AVF cyclotron. Inelastically scattered particles (alphas or protons) were detected by the high-resolution spectrometer Grand Raiden (GR) [15]. A schematic picture of the setup is shown in the Figure1. In the used forward scattering mode, the GR spectrometer was placed at the angles of 4.5 deg and 6.6 deg, for the case of the alpha and proton beams respectively. The Grand Raiden Forward-mode beam line (GRAF) was used to transport the beam to a beam dump placed at 20 m downstream of the target (see Figure 2). In this kind of experiment it is mandatory to stop the beam far from the target position and in a well-shielded beam dump, to decrease sufficiently the background in the gamma-ray coincidence measurements. The full solid angle of the Grand Raiden spectrometer is ~4 msr.

Two different target nuclei (^{90}Zr and ^{94}Zr) were studied using alpha and proton beams. The gamma-rays emitted following the de-excitation of these target nuclei were detected using both the clover type HPGe detectors of the CAGRA array [16] and the large volume $LaBr_3$:Ce scintillation detectors from the HECTOR+ array [17], placed in a 4π geometry around the target position (as can be seen in the inset of the Figure 1). The CAGRA array consisted of 12 clover detectors: 8 of them were placed at an angle of 90 deg. with respect to the beam direction, while the remaining 4 clovers were placed at backward (135 deg.) angles. The distance between the target and the front face of the HPGe detectors is 22.0 cm. The four $LaBr_3$:Ce scintillators (crystal dimensions 3.5"x8") of the HECTOR+ array were placed at forward angles (45 deg.) at a distance of 16.0 cm from the target to the front face of the detectors.

Figure 1. Picture of the Grand Raiden spectrometer in the west experimental hall of the RCNP, Osaka University. The inset show a picture of the gamma detectors placed around the target position: 12 HPGe clovers of the CAGRA array and 4 $LaBr_3$:Ce scintillation detectors of the HECTOR+ array.

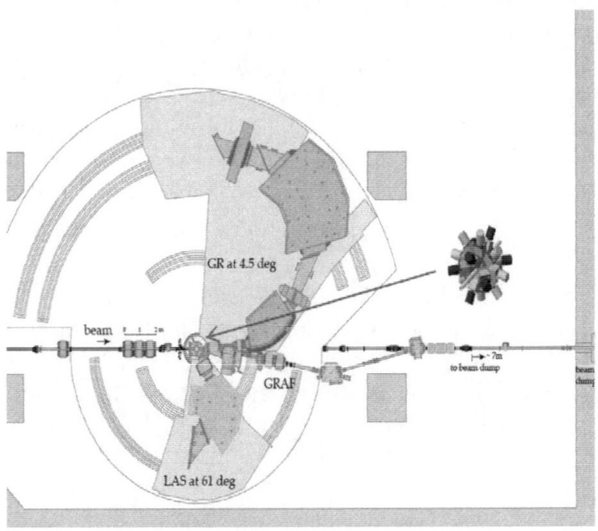

Figure 2. Schematic drawing of the Grand Raiden spectrometer used in forward scattering mode.

3. Preliminary Results

One important quantity in the analysis of the data is the correlation between E_x and E_{gamma}. E_x represents the excitation energy of the target nucleus, deduced from the measurement of the inelastically scattered particle (alpha or proton) in the Grand Raiden spectrometer. E_{gamma} is the energy of the gamma rays emitted following the de-excitation of the target nucleus, it is measured by the HPGe clover detectors of CAGRA or by the $LaBr_3$:Ce scintillation detectors of the HECTOR+ array. This correlation can be seen in the 2D-histogram shown in Figure 3. This refers to the case of the $^{90}Zr(\alpha,\alpha'\gamma)$ run (E_{beam}= 130 MeV) with E_{gamma} representing the gamma energy measured in the HPGe clover detectors. The events lying on the diagonal are those for which holds the condition $E_x=E_{gamma}$, they are associated to the direct decay to the ground state (this is shown in Figure 3 with the red dotted line). By applying different cuts in the the matrix E_x versus E_{gamma} it is possible to select events associated to different decay paths to the ground state. For example, the events lying along the black dotted diagonal line displayed in Figure 3 are associated to the decay to the first 2^+ excited state. In the matrix is also clearly visible the sudden decrease in the number of the events after an excitation energy E_x larger than the one neutron separation energy of ^{90}Zr (S_n =11.968 MeV) is reached, this is shown by the black vertical line in Figure 3. Background associated to random coincidences events has been subtracted in the matrix.

The E_x versus E_{gamma} 2D-histograms have been extracted for the four measurement runs associated to the different target-beam combinations used. In particular: for the case of the $^{90}Zr(\alpha,\alpha'\gamma)$ reaction with E_{beam} = 130 MeV; for the case of the $^{94}Zr(\alpha,\alpha'\gamma)$ reaction with E_{beam} = 130 MeV; for the case of the $^{90}Zr(p,p'\gamma)$ reaction with E_{beam} = 80 MeV and finally for the case of the $^{90}Zr(p,p'\gamma)$ reaction with E_{beam} = 80 MeV.

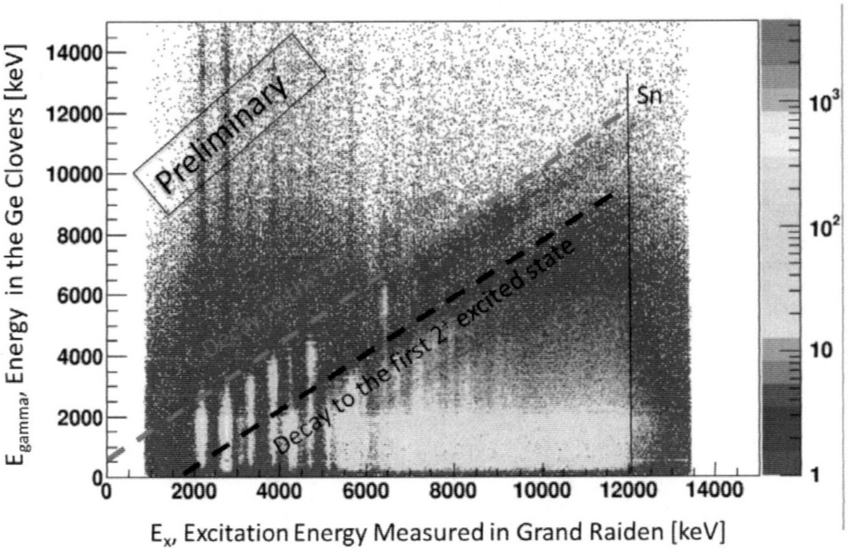

Figure 3. 2D-histogram showing the correlation between E_x (the excitation energy of the target nucleus) and E_{gamma} (energy of the gamma rays measured by the HPGe clover detectors of CAGRA) for the $^{90}Zr(\alpha,\alpha'\gamma)$ reaction at E_{beam} = 130 MeV.

These matrices show that states in the PDR energy region (below the neutron emission threshold) have been populated in $^{90,94}Zr$ for both the cases of alpha and proton beams.

The panel a) of figure 4 displays the comparison between the histograms of the excitation energy E_x (measured with Grand Raiden) of the ^{90}Zr target nuclei for the alpha inelastic reaction at E_{beam} = 130

MeV (red line histogram) and the proton inelastic scattering reaction at E_{beam} = 80 MeV (black line histogram). These histograms were obtained projecting on the X-axis the E_x vs E_{gamma} matrices and are associated to ground state gamma decay events. In fact in performing the projection only the events for which the following condition holds were considered: -40 keV < E_{diff} < 1100 keV; where E_{diff} corresponds to (E_x - E_{gamma}), that is the difference between the excitation energy of the target nuclei measured with Grand Raiden and the energy measured in the Ge gamma detectors. The two histograms are normalized in respect to the height of the prominent peak associated to the most intense 1^- state at 6425 keV, which is indicated in the figure. The panel b) of figure 4 show a similar comparison for the case of the measurements with the ^{94}Zr target nuclei. In this case the condition applied in performing the projection on the X-axis of the E_x vs E_{gamma} matrices is -30 keV < E_{diff} < 700 keV. The two histograms are normalized in respect to the height of the peak associated to the most intense 1^- state at 2846 keV, which is indicated in the figure. The neutron emission threshold is S_n=11.968 MeV for the ^{90}Zr and S_n=8.219 MeV for the ^{94}Zr. For both ^{90}Zr and ^{94}Zr there is clear similarity between the levels excited by protons and those excited by alpha particles in the lower energy part of the showed spectra (i.e. between 5.8 MeV and 7.6 MeV for the case of ^{90}Zr and between 2.8 keV and 5.0 MeV for the case of ^{94}Zr). In contrast, in the higher energy part the spectra appear to be significantly different. In particular for the case of ^{90}Zr the ratio of the counts in the higher energy region in the two spectra (alpha/proton) is 0.51 (in this case the higher energy region is considered from 7.6 MeV up to S_n=11.968 MeV). For the case of ^{94}Zr the same ratio (alpha/proton) is very different: 2.57 (in this case the higher energy region is considered from 5.0 MeV up to S_n=8.219 MeV).

These puzzling results are still preliminary and need further investigation, the data analysis is ongoing.

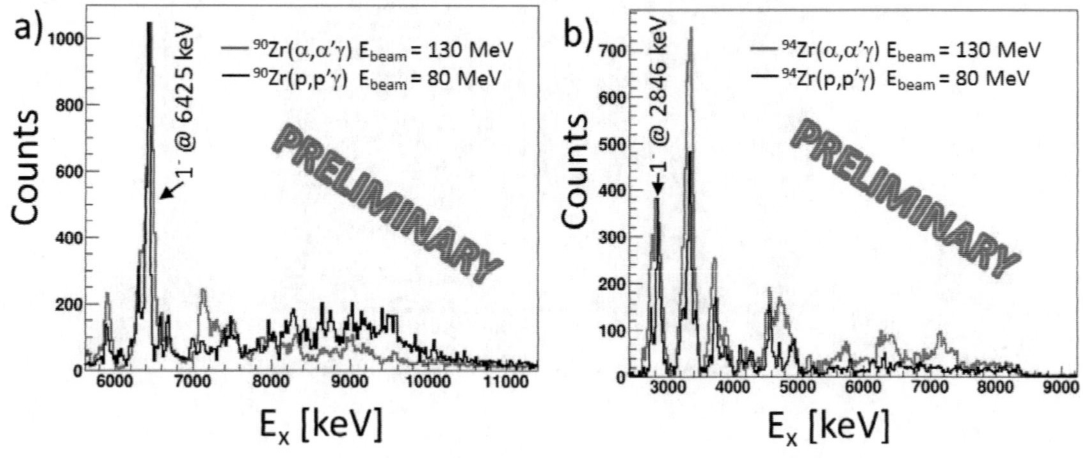

Figure 4. Panel a): Comparison between the histograms showing the excitation energy (measured with Grand Raiden) of the ^{90}Zr target nuclei for the alpha inelastic reaction at E_{beam} = 130 MeV (red line histogram) and the proton inelastic scattering reaction at E_{beam} = 80 MeV (black line histogram). These histograms are obtained as projections on the x-axis of the matrix E_x vs E_{gamma} (with the conditions described in the text) and are associated to ground state gamma decay events. The prominent peak associated to the most intense 1^- state at 6425 keV is indicated. Panel b): same comparison as in panel a) but this time the data are of the measurements with the ^{94}Zr target nucleus. The peak associated to the most intense 1^- state at 2846 keV is indicated.

Acknowledgements

It is acknowledged the German BMBF Grant No. 05P15RDFN1 and DFG under grant SFB 1245.

It is also acknowledged the support by the Extreme Light Infrastructure Nuclear Physics (ELI-NP) Phase II, a project co-financed by the Romanian Government and the European Union through the European Regional Development Fund – the Competitiveness Operational Programme (1/07.07.2016, COP, ID 1334).

References

[1] Savran D, Aumann T, and Zilges A 2013 *Prog. Part. Nucl. Phys.* **70** 210
[2] Bracco A, Crespi F C L, and Lanza E G 2015 *Eur. Phys. J A* **51** 99
[3] Roca-Maza X *et al.* 2012 *Phys. Rev. C* **85** 024601
[4] Vinas X *et al.* 2014 *Phys. Jour. A* **50** 27
[5] Paar N *et al.* 2007 *Rep. Prog. Phys.* **70** 691
[6] Goriely S, Khan E, and Samyn M 2004 *Nucl. Phys. A* **739** 331
[7] Savran D *et al.* 2006 *Phys. Rev. Lett.* **97** 172502
[8] Endres J *et al.* 2010 *Phys. Rev. Lett.* **105** 212503
[9] Crespi F C L *et al.* 2014 *Phys. Rev. Lett.* **113** 012501
[10] Pellegri L *et al.* 2014 *Phys. Lett. B* **738** 519
[11] Crespi F C L *et al.* 2015 *Phys. Rev. C* **91** 024323
[12] Krzysiek M *et al.* 2016 *Phys. Rev. C* **93** 044330
[13] Schwengner R *et al.* 2008 *Phys. Rev. C* **78** 064314
[14] Zweidinger M *et al.* 2012 *Proceedings of the XIX International School on Nuclear Physics, Neutron Physics and Applications, Journal of Physics: Conf. Series* **366** 012054
[15] Fujiwara M *et al.* 1999 *Nucl. Instrum. Meth. in Phys. Res. Sect. A* **422** 484
[16] http://www.rcnp.osaka-u.ac.jp/Divisions/np1-a/CAGRA/index.html
[17] Giaz A *et al.* 2013 *Nucl. Instrum. Meth. in Phys. Res. Sect. A* **729** 910

Integration of the GET electronics for the CHIMERA and FARCOS devices

E De Filippo[1], L Acosta[1,8], L Auditore[2,1], C Boiano[3], G Cardella[1], A Castoldi[3,5], M D'Andrea[1], S De Luca[2,1], F Favela[1], F Fichera[1], N Giudice[6,1], B Gnoffo[6,1], A Grimaldi[1], C Guazzoni[3,5], G Lanzalone[4,7], F Librizzi[1], P Litrico[4], C Maiolino[4], S Maffesanti[3,5], NS Martorana[6,4], A Pagano[1], EV Pagano[4], M Papa[1], T Parsani[3,5], G Passaro[4], S Pirrone[1], G Politi[6,1], F Previdi[3,5], L Quattrocchi[2,1], F Rizzo[6,4], P Russotto[4], G Saccà[1], G Salemi[6,1], D Sciliberto[1], A Trifirò[2,1] and M Trimarchi[2,1]

[1] INFN, Sezione di Catania, Italy
[2] Dipartimento di Scienze MITF, Università di Messina, Italy
[3] INFN, Sezione di Milano, Italy
[4] INFN, Laboratori Nazionali del Sud, Catania, Italy
[5] Dip. di Elettronica, Informazione e Bioingegneria, Politecnico di Milano, Italy
[6] Dip. di Fisica e Astronomia, Università di Catania, Italy
[7] Università di Enna, "Kore", Italy
[8] Instituto de Física, Universidad Nacional Autónoma de México, México

E-mail: defilippo@ct.infn.it

Abstract. A new front-end based on digital GET electronics has been adopted for the readout of the CsI(Tl) detectors of the CHIMERA 4π multi-detector and for the new modular Femtoscopy Array for Correlation and Spectroscopy (FARCOS). It is expected that the coupling of CHIMERA with the FARCOS array, featuring high angular and energy resolution, and the adoption of the new digital electronics will be well suited for improving specific future data analysis, with the full shape storage of the signals, in the field of heavy ion reactions with stable and exotic beams around the Fermi energies domain. Integration of the GET electronics with CHIMERA and FARCOS devices and with the local analog data acquisition will be briefly discussed. We present some results from previous experimental tests and from the first in-beam experiment (Hoyle-Gamma) with the coupled GET+CHIMERA data acquisition.

1. Introduction

The CHIMERA 4π detector, originally conceived for multi-fragmentation and reaction dynamics studies in the Fermi energy domain [1,2], has passed different upgrades, in agreement with new experimental and physical needs. The most recent one was the Silicon Pulse Shape upgrade (PSD) [3], aiming to measure the rise time of the signals for the charged particles stopping in the silicon detector (the first detection layer of the CHIMERA telescope). In this way the threshold for simultaneous mass and charge identification for light fragments was extremely lowered. This last aspect is particularly important when the isospin degree of freedom is investigated in nuclear reactions in order to study quantities related to the Equation Of State (EOS) of asymmetric nuclear matter and in-medium nuclear effective interactions [4,5].

The availability of radioactive beams at LNS [6], that has also opened the use of CHIMERA detector to nuclear structure and clustering studies, requires new constraint in the particles detection, like the capabilities to detect with good efficiency, together with light charged particles, γ-rays and/or neutrons. Recently, the capability of the CHIMERA CsI(Tl) detectors for γ-rays identification was exploited in a study upon the 4.44 MeV excited level of ^{12}C [7] and in the search of the isoscalar excitation of the Pygmy resonance in ^{68}Ni [8].

Amongst other observables, two and multi-particle correlations are important tools to characterize the time scale and shape of emission sources in the dynamical evolution of heavy ion collisions. But this imposes both high energy and angular resolutions in particle detection. FARCOS is an ancillary and compact multi-detector with high angular granularity and energy resolution for the detection of light charged particles (LCP) and Intermediate Mass Fragments (IMF) designed for multi-particle correlation measurements and "femtoscopy" studies [9–11]. The FARCOS array is constituted by 20 telescopes in the final project. Each telescope is composed by three detection stages: the first ΔE is a 300 μm thick DSSSD silicon strip detector with 32x32 strips; the second is a DSSSD, 1500 μm thick with 32x32 strips; the final stage is constituted by 4 CsI(Tl) scintillators, each one of 6 cm in length.

A single FARCOS telescope has to handle at least 196 readout channels taking into account a double amplification dynamics, limited to front strips. Thus the final FARCOS array needs to handle around 4000 channels. Moreover the electronic front-end of the 1192 CHIMERA CsI(Tl) scintillators, based on VME readout [12], was becoming obsolescent in the last ten years of successful runs. For these reasons a generic and scalable electronic system, covering digitalization, signals readout, synchronization and triggering based on the GET electronics (General Electronics for TPCs) [13, 14] has been adopted. The integration of this system has required the development of a new first stage front-end for FARCOS, based on CMOS new preamplifiers for both DSSSD Silicon [15] and CsI(Tl) detector [16], integrated in a single and compact ASIC board, and the design of a new dual-gain module to fit with the wide dynamical energy ranges expected for the CHIMERA CsI(Tl) and FARCOS Silicon detectors. Indeed a careful redesign of CHIMERA data acquisition was needed in order to couple the GET digital acquisition with Chimera Silicon detectors readout handled by the analog acquisition on VME bus.

In the following, only few selected aspects of the upgrade project will be shown, with particular emphasis on qualification of the GET electronics in particle identifications and data acquisitions integration.

2. Qualification of the GET electronics with CHIMERA detectors

The GET electronics, based on the integrated circuit AGET chip, originally designed for Time Projection Chambers (TPCs) and active targets devices has been adapted for CsI(Tl) scintillators and Silicon strip detectors of CHIMERA and FARCOS arrays. In particular the main idea was to by-pass the pre-amplification stage embedded in the AGET chip in order to use the existing CHIMERA pre-amplifiers operating under vacuum [17] and to design a new CMOS front-end in the Silicons and Csi(Tl) for the FARCOS array [15, 16]. Signals coming from the pre-amplifiers are interfaced to the GET system through a Dual Gain Module. This last is a multichannel digitally programmable gain amplifier, two ways splitter and level adapter, interfaced to the AGET inputs. In this way the large dynamic range of more than 10^3 required by CHIMERA and FARCOS devices can be achieved.

The GET electronics is described in detail in Ref. [14] and references therein. Very briefly, the AGET chip integrates 64 channels with a pre-amplifier, an analog filter (shaper) and leading-edge discriminator. Four further channels, Fixed-Pattern Noise (FPN), are used to evaluate the intrinsic noise level and baseline shape of the channels. The filtered signals are sampled trough storing in a analog memory based on a Switched Capacitor Array (SCA) structure (512-cell

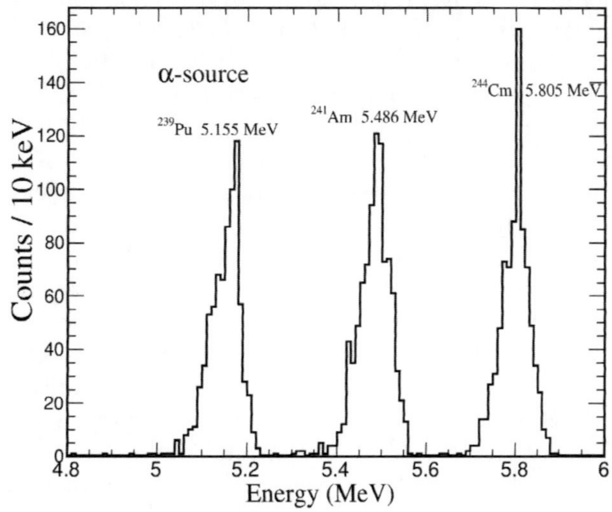

Figure 1. Energy spectrum of a three peaks α source relative to a single strip (front side) of a 1500 μm thick DSSSD strip detector of a FARCOS telescope, obtained by 50 MS/s digitized pulses.

deep) and codified by a 12 bit ADC. A sampling rate up to 100 MS/s can be set-up. Four AGET chips are arranged in a card called AsAd (ASIC and ADC) that acts as a front-end for the GET system. Up to 4 AsAd board (1024 channel) can be connected to the CoBo board that manages the slow control of the AGET chips and the readout of the AsAd(s) data at a maximum bandwidth of 1.2 Gb/s for each AsAd. CoBo boards are housed in a μTCA (Micro Telecom Computing Architecture) crate together with the trigger box, "MuTanT". A μTCA crate can handle up to 11 CoBo boards and it handles also the data sending to a computer farm for data analysis and storage through the Ethernet link backplane at 10 Gb/s maximum bandwidth.

In order to characterize the GET electronics for CHIMERA and FARCOS devices an extensive testing phase has been performed during the hardware and software upgrade developments. We show in this section some selected results of this preliminary phase. Fig. 1 shows the energy spectrum of a mixed nuclei (three peaks) α source irradiating under vacuum a DSSSD strip detector, 1500 μm thick, 64x64 mm^2 dimension, of a FARCOS telescope. The spectrum is relative to one strip of the detector front side. In this test the detector was coupled to a commercial (Mesytec) charge pre-amplifier ($\sim$ 13 mV/MeV) followed by a Dual Gain module (gain = 2) interfaced to the AsAd input. The digitized output is filtered with the AGET front-end analog filter with a peaking time of 1 μs and digitized at 50 MS/s in the SCA to be codified by a 12 bit ADC. The output waveforms are software analyzed (FPN subtraction, baseline restore, digital filtering with a Finite impulse Response (FIR) triangular filter, see Refs. [18–21] for an overview of typical digital signal processing with CHIMERA detectors) in order to perform on-line and off-line computation on the pulse waveform. Fig. 1 presents the spectrum distribution of the signal pulse height and shows a resolution of 60-70 keV (FWHM) that is on average when operating on this kind of detectors by using GET electronics at 50 MS/s and commercial pre-amplifiers.

We have performed different tests under Tandem and Cyclotron beams at LNS in order to check the GET electronic in standalone mode (not coupled with CHIMERA data acquisition)

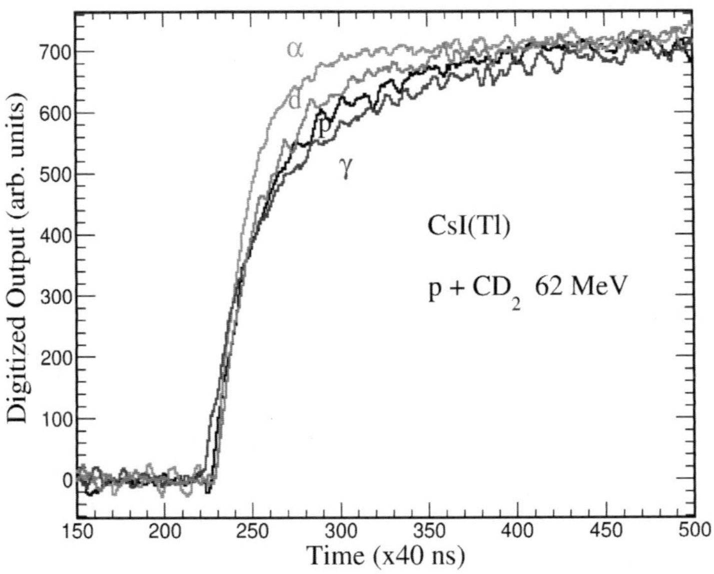

Figure 2. Baseline subtracted output waveforms (unfiltered), collected at 25 MS/s, for different light charged particles and γ-rays detected by a CHIMERA CsI(Tl) of ring 9 (R9-37E, θ=28.5^o) in the reaction p+CD_2 (deuterated polyethylene target) at 62 MeV (see text for details).

with the CHIMERA telescopes, for both Si and CsI(Tl) detectors. In one of the performed tests we used a proton (62 MeV) beam and some telescopes of ring 9 (θ=28.5^o) by using the standard CHIMERA pre-amplifiers for Silicons and CsI(Tl) detectors, these last with photodiode readout of the light output. Signals from Si and CsI(Tl) were both digitized at the relative low frequency of 25 MS/s (40 ns step) in this experiment, mostly devoted to CsI(Tl) response. The CsI(Tl) signal was shaped with the AGET Sallen-Key filter with a small peaking time (220ns), so preserving the rise time information. CsI(Tl) light particles and γ identification was obtained by calculating the pulse height (PH) vs. the Rise Time of the signals read at the output of the pre-amplifiers. This identification technique has be shown to be equivalent to the conventional Fast-Slow components scatter plot of the light emission [18]. Fig. 2 shows a sample of signals of different particles (p,d,α, γ-rays), identified in the PH vs Rise-Time scatter plot and having approximately the same pulse heights. The waveform outputs shown in the figure were baseline restored; no software FIR filter was applied. This collections of traces nicely show how faster is the light collection for α particles in CsI(Tl) with respect to deuterons, protons and γ-rays respectively, giving the possibility of an off-line optimizations of the identification procedures.

We have performed tests under beam with CHIMERA detectors and GET in standalone trigger mode during the CLIR (Clustering in Light Ions Reactions) experiment [22,23] with the FRIBs (in Flight Radioactive Ions BeamS) facility at LNS. A fragmentation cocktail beam was produced starting from a 55 MeV/nucleon ^{18}O primary beam impinging on a 1.5 mm thick ^{9}Be target for fragments production. The cocktail beams, collected after the fragments separator, are sent to the reaction target inside the vacuum chamber. The CHIMERA tagging system can select, event-by-event, each incoming isotope [24]. Ions like ^{10,11}Be, ^{13}B and ^{16}C were produced with on average 10^4-10^5 pps rate depending on selected ion.

In this experiment some telescopes of the ring 2 of CHIMERA (θ = 4°) were connected

IOP Conf. Series: Journal of Physics: Conf. Series **1014** (2018) 012003 doi:10.1088/1742-6596/1014/1/012003

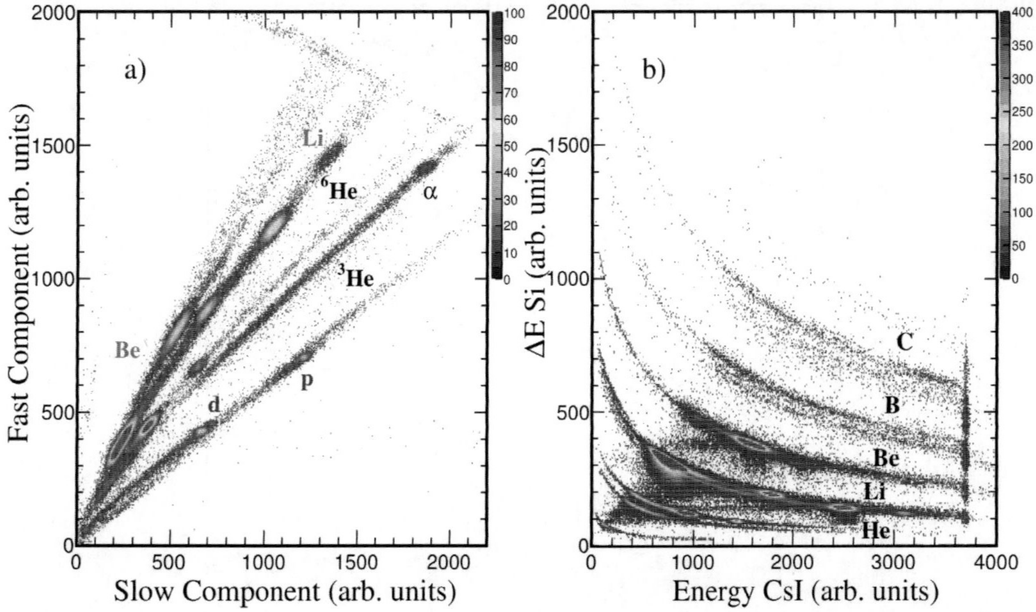

Figure 3. Data obtained with GET electronics and data acquisition for reaction products on a telescope of ring 2 (R2-18E, θ=4°) from a cocktail beam on a thin polyethylene $(CH_2)_n$ target. a) Fast vs. Slow component identification in CsI(Tl) detector. b) Particle identification from the ΔE energy loss in the 300 μm Silicon detector vs the residual energy in CsI(Tl) detector. In these plot bumps are of course relative to the elastic peaks of the cocktail beams

to the GET electronics with the purpose to compare results with those from standard analog acquisition. Fig. 3 show results of particles identification obtained with the GET system. A 50 μm thick polyethylene $(CH_2)_n$ target was used in the shown data. We tested in different runs a sampling rate of 25 or 50 MS/s for both Silicon and CsI(Tl) detectors (when using a single CoBo a unique sampling rate can be selected for all channels). Fig. 3 refers to the 25 MS/s sampling rate setting. CsI(Tl) fast and slow components, CsI(Tl) and Si pulse heights and rise times were calculated on the digitized outputs. For the determination of the Fast and Slow components, shown in Fig. 3a), we used the gates definition method described in Fig. 1 of Ref. [18], best suited for signals coming directly from a pre-amplifier output. In this method the Fast (Slow component) is obtained as the pulse value interval $(H_2 - H_1)$ corresponding to a given time window gate, $(t_2 - t_1)$. The time gate are 800 ns and between 1.4μs and 10μs respectively for the Fast and Slow component in Fig. 3a), calculated respect to the starting point (10% of the pulse height) of the signal. Fig. 3b) shows the Silicon Pulse Height distribution (Si energy loss) as function of the Csi(Tl) pulse height distribution (residual energy). It is interesting to note that, despite the poor sampling rate for a silicon detector, a good resolution is obtained for isotopic separation up to Carbon. This result for detectors of ring 2 in Chimera, that have just a 2 mV/MeV gain silicon pre-amplifier, is generally not reached with standard analog acquisition also in this range of low charge ions. Indeed a saturation in the CsI(Tl) light output is clearly seen in both bidimensional plots. Thus we argue that recovering the full energy range dynamic requires the use of the Dual Gain module.

3. Integrations of the data acquisitions

A part of the CHIMERA 4π detector, limited to CsI(Tl) detectors, will use the GET front-end electronics as well the FARCOS telescopes. On the other hand the current electronic and readout front-end for Silicon detectors are expected to continue to be used. The actual electronic chain for CHIMERA Silicon detectors has, as final element, the analog codifiers (QDC and TDC, respectively Charge and Time to Digital convertes) housed in the VME bus. The total number of GET channels for the CHIMERA + FARCOS (20 telescopes) devices is around 6000. This number corresponds to about 24 AsAd distributed up to 8 CoBo modules. Thus, a single μTCA crate can handle the whole configuration. The GET data acquisition (based on CoBo readout modules and the MuTanT trigger in the μTCA bus) is described in detail in Refs. [13, 14].

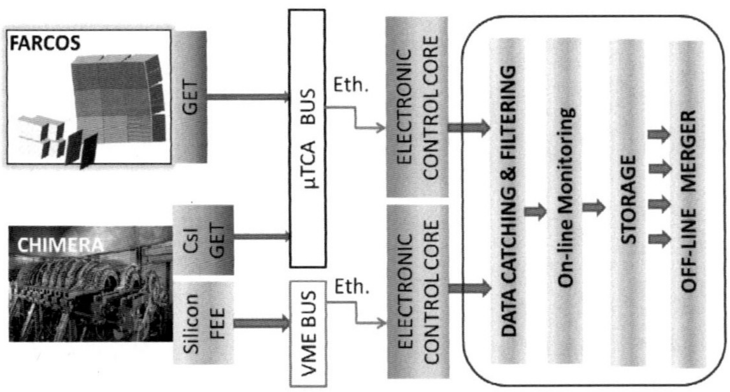

Figure 4. Schematic view of CHIMERA CsI(Tl) (GET), CHIMERA Silicon (VME DAQ) and FARCOS DAQ (GET) coupling. See text for details.

Fig. 4 shows schematically the data flow from the front-end electronics to the data storage. Initial tasks are the data collection from the readout server (for the VME bus) and the CoBo boards (for GET). Intermediate tasks are data filtering and on-line monitoring. Data storage and merging of the data belonging to the same event (event building) are the final tasks. We use, in order to manage all these tasks (excluding the data merging) the distributed acquisition framework NARVAL [25]. A NARVAL process, called "actor" is a task serving a specific function in the data flow. Both CHIMERA data acquisition and the GET one are supervised and coordinated by the NARVAL system.

The DAQ coupling is related on the use of a common trigger configuration. In this case the data merger among different acquisitions can be easily based on Timestamps and/or event number that has to be the same for a each event generated by the same trigger in different acquisition systems. The MuTanT trigger handles a Global Master Clock (GMC) at 100 MHz synchronizing all CoBo modules belonging to same crate and the 14 bit Timestamp recorded in each single CoBo. The MuTanT has three trigger levels (L0, L1, L2). L0 is an external trigger (for example generated by CHIMERA silicon detectors in our case); L1 is generated by the internal total multiplicity, derived by the sum of the individual channel multiplicities in all AsAd devices; L2 is a user software implemented trigger. Unfortunately there is no compatible Timestamp device on the CHIMERA VME bus. Therefore the merging between the VME events and the GET ones is handled by using the event counter that essentially acts as a trigger number counter both in VME and GET systems. Note that the method of common trigger and event counter can be successfully used only if the DAQ dead time is shared between the two different data acquisitions, handling in a proper way the busy signal from the different DAQ and front-end systems.

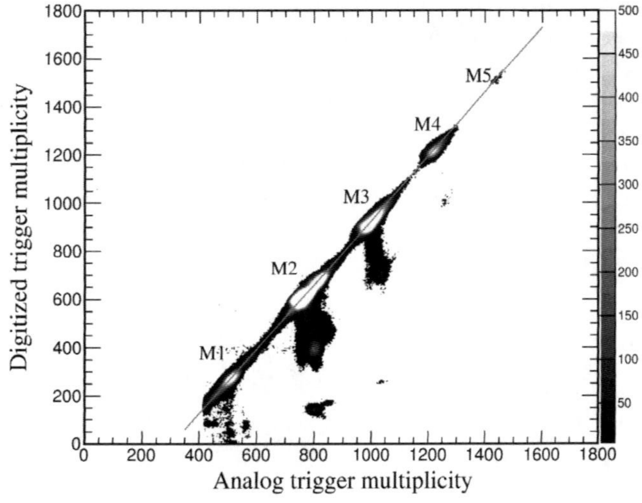

Figure 5. 2D plot showing the correlation between the analog sum multiplicity signal distribution generated by the CHIMERA Silicon detectors and sent: in Y-axis to a channel of the GET system, to be analyzed by the digital DAQ; in X-axis to a QDC channel in the VME bus to be integrated by the analog DAQ. Correlated events with the same event number should stay along a diagonal line.

In the event builder the fact that two events coming from the VME and GET buffer respectively have the same event counter is only a necessary condition but not sufficient in order to affirm that the two events are correlated. In order to monitor, event-by-event, the data correlation we use the procedure described in Fig. 5. The multiplicity signal (analog sum signal) generated by the CHIMERA constant fraction discriminators for Silicon detectors is integrated in a QDC channel of the VME DAQ. The same signal is splitted, sent to an AsAd channel of the GET electronics and analyzed in pulse height for each valid trigger. The correlation between the two distributions after the event building have to lay in a diagonal line (as shown in Fig. 5). Conversely it will show a random behaviour for runs with uncorrelated events.

In order to perform the event building a new application called "DataMerger" has been developed [26]. The DataMerger manages data collected by CoBo(s) in the GET digital acquisition and VME analog DAQ and merges data belonging to the same trigger/event on the basis of the timestamp and/or event number. The DataMerger is also an on-line or off-line program for data analysis and signal processing. In fact, the DataMerger does not work directly upon raw data frames in order to do merging, but at the level of pre-analysed data collected for each single partial event. In this way the final results of the merging action are physical raw data that are ready to be stored in Root trees [27] or spectra for on-line and off-line monitoring. Fig. 6 shows the logical data flow of the application. The program generally collects data stored on disk on several files created by each Narval Storage Actor that are related to each CoBo present in the data acquisition and to VME data flow, because no merging is performed at DAQ level. Of course each stored file contains only data related to the single CoBo (or VME). Note that in this way the data storage is faster because it is distributed on different tasks and files equal to the number of different CoBo modules (plus the VME Daq buffers) used.

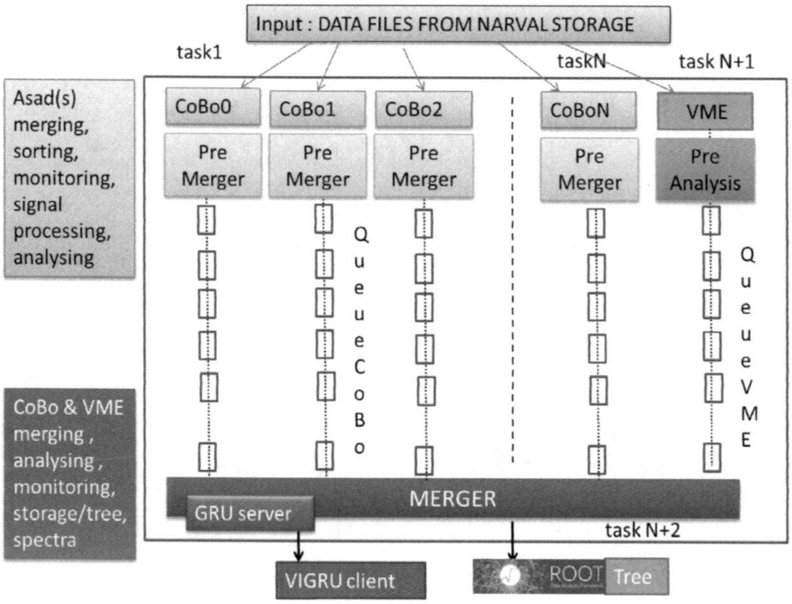

Figure 6. Data flow of the Merger application for CHIMERA-GET event building. Computation of digitized waveforms and AsAd frames merging (pre-merger) belonging to a given CoBo are done by single tasks concurrency running on separate threads. ViGRU [28] is a client for on-line visualization of Root histograms.

4. The Hoyle-Gamma experiment

The GET integration in CHIMERA detectors has been validated in November 2017 during the Hoyle-Gamma experiment at LNS. The goal of this experiment was to search for the γ-decay branching ratio of the 7.65 MeV Hoyle state and the rare γ-decay modes in ^{12}C, in particular, the 3_1^- state at E_x = 9.64 MeV [29–31], whose γ-decay probability is very small and poorly known.

In the experiment a ^{4}He (64 MeV) ions beam delivered by the Superconducting Cyclotron at INFN-LNS was used, impinging on a thin ^{12}C target. The main idea of this experiment was to exploit the 4π coverage of the CHIMERA detector to detect and identify, by using the CsI(Tl) scintillators, the γ-ray cascade relative to the decay of the 7.65 Hoyle and the 9.64 excited levels in ^{12}C and to detect in coincidence the inelastic scattered α particles and the ^{12}C recoil nucleus [7,32]. The kinematical coincidence technique should permit also to extract meaningful angular distributions for γ-rays as shown in Ref. [7].

In the experiment the GET digital electronics was used for the readout of the all CsI(Tl) scintillators of the CHIMERA backward part (the Sphere, $30^o < \theta < 176^0$). Dual Gain modules were used to split all CsI signals in two different gains in particular in order to improve the signal to noise ratio for the low energies γ-rays. Signals coming from the Sphere pre-amplifiers were splitted by 8 Dual Gain modules (around 1000 channels) and sent to 4 AsAd distributed over 2 CoBo modules for data readout. A remaining 2 AsAd connected to a third CoBo were used for control and monitoring purposes. All signals were sampled at 50 MS/s. The remaining CsI(Tl) of the CHIMERA forward part and the Silicon detectors were readout with the usual electronic front-end and analog readout in the VME bus. Events were triggered requiring a multiplicity M$\geq$2 in the Silicon detectors. The multiplicity signal was sent to the L0 input (external trigger) of the MuTanT trigger box. The acquisition rate was around 1 kHz and all the sampled signals

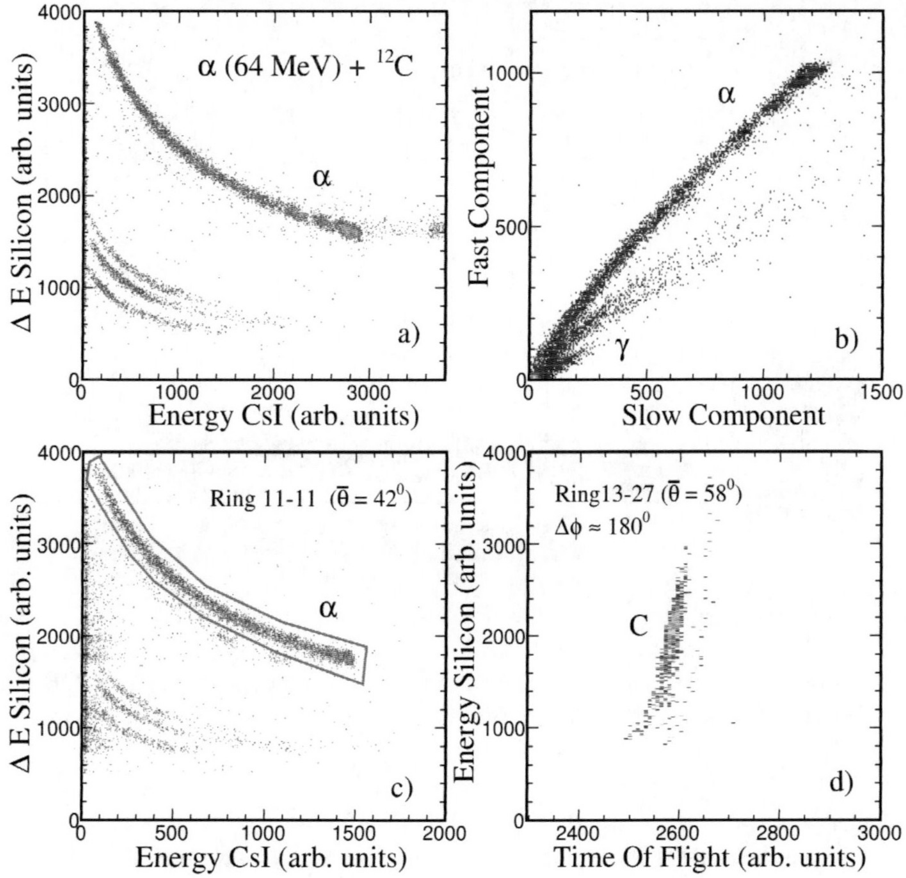

Figure 7. a) ΔE-E bidimensional plot for a telescope of ring 10 in the CHIMERA sphere; b) the corresponding Fast-Slow component of light emission in the CsI(Tl) 2D plot; c) ΔE-E bidimensional plot for a telescope of ring 11. d) Energy vs Time-of-Flight 2D plot obtained imposing a kinematical coincidence between the α selected by a cut in Fig. 7c) and a telescope of ring 13 with a $\Delta\phi \approx 180^o$, showing the Carbon recoil.

from CsI(Tl) detectors were stored on disk. A maximum total rate of 150 MB/s of stored data was reached during the acquisition.

The data analysis requires as first step an accurate calibration of CsI(Tl) and Silicon detectors that is not yet available. We show a sample of first results in Fig. 7. Fig. 7a) shows the ΔE energy loss in Silicon vs. residual energy in CsI for a CHIMERA telescope of ring 10 ($\theta = 34^o$). In this bidimensional plot the ΔE is obtained by the analog front-end and VME DAQ while the x-axis (residual energy) is obtained by the pulse height distribution of the relative CsI(Tl) channel in the GET digital electronic. Fig. 7b) shows the corresponding Fast vs. the Slow component bidimensional plot giving γ and light particles identification in CsI(Tl) obtained by pulse shape analysis of the sampled data. Finally we tested the coincidences of (in)-elastically scattered α particles with ^{12}C recoils that is one of the task of the experiment. Selecting alpha particles in a telescope of ring 11 (Fig. 7c)) we expect the maximum yield for the ^{12}C recoil in a telescope of ring 13 with an azimuthal angle difference between the two telescopes of $\Delta\phi \approx 180^o$.

The Carbon recoil stops in the Silicon detector, thus can be only detected by Time-of-Flight technique. This is well shown in Fig. 7d) where the Silicon energy vs. the Time-of-Flight (determined respect the Cyclotron Radio-Frequency) is plotted.

The examples shown in Fig. 7 give a first view of the general correctness of the adopted procedures in the upgrade phase and during the experiment and they are a good start for the prosecution of the data analysis.

5. Conclusion

Following the building of a first FARCOS module demonstrator and the approval for the construction of 5 FARCOS modules (20 telescopes) we have adopted a compact electronic front end based on the design of new ASIC preamplifiers for silicon strips and FARCOS CsI(Tl) and the GET electronics for digitalization and data readout. The front-end for CsI(Tl) CHIMERA detectors has been upgraded by using GET electronic as well.

A CHIMERA-GET hybrid data acquisition and the new Dual Gain modules designed at INFN-Catania have been successfully tested during the Hoyle-Gamma experiment.

The main goal for the next future is that the FARCOS ancillary detector, with its high energy and angular resolution, can enhance the physics that can be studied with the CHIMERA array (including two and multi-particles correlations) both with stable and radioactive beams, in the field of dynamics of heavy ion collisions and spectroscopy of light nuclei in the vicinity of drip lines. In view of this goal, studies to include also the neutron detection in CHIMERA and FARCOS have been recently started [33] in order to provide unique information about nuclear dynamics and spectroscopy working with exotic and stable beams.

References

[1] Pagano A 2012 *Nuclear Physics News* **22,1** 25
[2] De Filippo E and Pagano A 2014 *Eur. Phys. J. A* **50** 32
[3] Alderighi M et al. 2005 *IEEE Trans. Nucl. Sci.* **52** 1624
[4] Horowitz C et al. 2014 *J. Phys. G* **41** 093001
[5] Oertel M, Hempel M, Klahn T and Typel S 2017 *Rev. Mod. Phys.* **89** 015007
[6] Russotto P et al. 2018 Contribution to this conference
[7] Cardella G et al. 2015 *Nucl. Instrum. Meth. Phys. Res. A* **799** 64
[8] Martorana N S et al. 2018 *submitted to Phys. Lett. B*
[9] FARCOS Technical Design Report 2015, available on-line at: `https://drive.google.com/file/d/0B5CgGWz8LpO0c3pGTWdOcDBoWFE`
[10] Pagano E V et al. 2016 *EPJ Web of Conf.* **117** 10008
[11] Acosta L et al. 2016 *J. Phys.: Conf. Series* **730** 012001
[12] Alderighi M et al. 2002 *Nucl. Instrum. Meth. Phys. Res. A* **489** 257
[13] Giovinazzo J et al. 2016 *Nucl. Instrum. Meth. Phys. Res. A* **840** 15
[14] Pollacco E C et al. 2018 *Nucl. Instr. Meth. Phys. Res. A* **887** 81
[15] Castoldi A, Guazzoni C and Parsani T 2014 *Proc. IEEE Nucl. Sci. Symp. Med. Imag. Conference, Seattle Wa, Nov. 2014* 1-5
[16] Castoldi A, Guazzoni C and Parsani T 2017 *IEEE Trans. Nucl. Sci.* **64** 2678
[17] Bassini R, Boiano C, Pagano A, Pullia A 2004 *IEEE Trans. Nucl. Sci.* **51** 115
[18] Acosta L et al. 2013 *IEEE Trans. Nucl. Sci.* **60** 284
[19] Amorini F et al. 2007 *IEEE Trans. Nucl. Sci.* **54** 208
[20] Amorini F et al. 2012 *IEEE Trans. Nucl. Sci.* **59** 1772
[21] De Luca S 2016 Doctoral Thesis, Catania University
[22] Dell'Aquila D et al. 2016 *Phys. Rev. C* **93** 024611
[23] Dell'Aquila D et al. 2017 *Acta Phys. Polonica B* **48** 499
[24] Lombardo I et al. 2011 *Nucl. Phys. (Proc. Suppl.) B* **215** 272
[25] X. Grave et al 2005 *Proc. 14th IEEE Real Time Conference* 119
[26] De Filippo E 2017 DataMerger: an object oriented Merger for CHIMERA-GET data acquisition: User's Guide.
[27] Brun R and Rademakers F 1997 *Nucl. Inst. and Meth. in Phys. Res. A* **389** 81
[28] Legeard L *GRU/ViGRU documentation*: `http://wiki.ganil.fr/gap/wiki/Documentation/Software/Gru`

[29] Tsumura M et al. 2014 *Jour. of Phys.: Conf. Series* **569** 012051
[30] Alsharani B et al. 2013 *EPJ Web of Conferences* **63** 01022
[31] Cardella G et al. 2017 *EPJ Web of Conferences* **165** 01009
[32] Cardella G et al. 2016 Measurement of the γ-decay branching ratio of the Hoyle and first excited 3^- levels of ^{12}C. *Experiment proposal PAC-LNS/2016*
[33] Pagano E V et al. 2018 *Nucl. Instrum. Meth. Phys. Res. A* **889** 83

Study on reaction mechanism by analysis of kinetic energy spectra of light particles and formation of final products

G Giardina[1], G Mandaglio[2,3], A K Nasirov[4,5], A Anastasi[1], F Curciarello[6] and G Fazio[1]

[1] Dipartimento di Scienze Matematiche e Informatiche, Scienze Fisiche e Scienze della Terra, University of Messina, Messina, Italy
[2] Dipartimento di Scienze Chimiche, Biologiche, Farmaceutiche ed Ambientali, University of Messina, Messina, Italy
[3] INFN Sezione di Catania, Catania, Italy
[4] JINR - Bogoliubov Laboratory of Theoretical Physics, Dubna, Russia
[5] National University of Uzbekistan, Tashkent, Uzbekistan
[6] INFN Laboratori Nazionali di Frascati, Frascati, Italy

Abstract. The sensitivity of reaction mechanism in the formation of compound nucleus (CN) by the analysis of kinetic energy spectra of light particles and of reaction products are shown. The dependence of the $P_{\rm CN}$ fusion probability of reactants and $W_{\rm sur}$ survival probability of CN against fission at its deexcitation on the mass and charge symmetries in the entrance channel of heavy-ion collisions, as well as on the neutron numbers is discussed. The possibility of conducting a complex program of investigations of the complete fusion by reliable ways depends on the detailed and refined methods of experimental and theoretical analyses.

1. Introduction

The study of nuclear reactions in heavy-ion collisions continues to be of great interest in the scientific community to better understand the reaction dynamics from the stage of capture of the projectile by target up to the formation of the final products. Nevertheless, in literature there are still present relevant unclear discrepancies between experimental and theoretical results. Uncertainties also exist among various theoretical results and incomprehension also among various experimental results. The differences between theoretical results by different models are related to the assumptions made in the procedures of theoretical calculations and use of simplified phenomenological models unsuitable to describe the reaction dynamics. In fact, the use of free parameters can lead to an apparent acceptable agreement between the calculated results and data but actually these results can not prove the effectiveness of the procedures used to obtain the experimental results. Analogously, the procedure of obtaining the best values of free parameters used in the phenomenological model which leads to results of calculation in good agreement with the obtained experimental results can not demonstrate by a clear and unambiguous way the understanding of the reaction dynamics. The main reason for the differences in the experimental fusion and capture cross sections is related to the ambiguity of the procedures at the separation of the events corresponding to deep-inelastic collisions (DICs), quasifission (QF) and fusion-fission (FIS) processes. The quasifission is the decay of the dinuclear system (DNS) into two fragments without formation of CN. There is still no definite understanding the nature of full momentum transfer reactions at the experimental analysis of the deep-inelastic collisions and quasifission events to estimate capture cross sections. The overlap of the mass and/or angular distributions of the quasifission and fusion-fission products causes ambiguity in the estimation of the experimental fusion cross sections. The P_{CN} fusion probability of reactants in the heavy ion collisions is estimated as a ratio of the complete fusion ($\sigma_{\rm fus}$) and capture ($\sigma_{\rm cap}$) cross sections:

$$P_{\rm CN} = \frac{\sigma_{\rm fus}}{\sigma_{\rm cap}} = \frac{\sigma_{\rm fus}}{\sigma_{\rm fus} + \sigma_{\rm qfis}}. \quad (1)$$

The theoretical value of the capture cross section is determined by the estimation of the range of the orbital angular momentum leading to the full momentum transfer in the entrance channel of collision. The evolution of the excited dinuclear system (DNS) formation can lead to complete fusion in competition with quasifission. The details of this model are present in many of our papers (see, for example, [1–11]).

The model takes into account the dependence of the capture cross section on the range of the orientation angles of nuclei at the initial stage of nuclear collision, the mass asymmetry parameter of reactants in the entrance channel, the considered $E_{c.m}$ energy range for the investigated reaction. The orbital angular momentum range is needfull to consider the evolution of each reaction from the contact of reactants to the compound nucleus formation, until to obtaining the final products of deexcitation of CN.

It is obvious the differences between the values of $P_{\rm CN}$ extracted from the measured data of the capture and complete fusion events depend on how are correctly estimated capture and fusion cross sections from the measured data. Therefore, the reliable experimental determinations of $P_{\rm CN}$ and consequently the understanding the entrance channel effect on the $P_{\rm CN}$ values are strongly related to the choice of assumptions for the data analysis.

The competition between the complete fusion of nuclei in DNS and quasifission (decay of DNS into two fragments) processes decreases the value of the fusion cross section [12, 14, 15]:

$$\sigma_{\rm fus}(E_{\rm c.m.}) = \sum_{\ell=0}^{\ell_d(E_{\rm c.m.})} (2\ell+1)\sigma_{\rm cap}(E_{\rm c.m.},\ell)P_{\rm CN}(E_{\rm c.m.},\ell). \quad (2)$$

The maximum value of ℓ leading to capture $\ell_d(E_{c.m.})$ depends on the beam energy and it is calculated by the solution of the radial motion equations (see ref. [15]). Since the capture cross section is equal to the sum of the complete fusion and quasifission cross sections, $\sigma_{\rm cap} = \sigma_{\rm fus} + \sigma_{\rm qfis}$, the quasifission cross section is calculated by the expression

$$\sigma_{\rm qfis}(E_{\rm c.m.}) = \sum_{\ell=0}^{\ell_d}(2\ell+1)\sigma_{\rm cap}(E_{\rm c.m.},\ell)(1 - P_{\rm CN}(E_{\rm c.m.},\ell)). \quad (3)$$

It should be stressed that quasifission of dinuclear system can take place at all angular momentum values from $\ell = 0$ to ℓ_d. Another binary process which leads to the formation of two fragments similar to those of fusion-fission and quasifission is the fast fission (FF). The fast fission occurs only if there is not a fission barrier for the being formed compound nucleus due to the large values of the angular momentum, $\ell > \ell_f$. According to the rotating liquid drop model (see [16]) a rotating nucleus with the angular momentum ℓ_f breaks down immediately. Therefore, FF is determined as the disintegration of the fast rotating mononucleus into two fragments, through DNS survives quasifission to be transformed into CN. In the case of the superheavy nuclei, the fission barrier providing their stability against fission appears only due to shell effects in their binding energy because there is no barrier connected with the liquid-drop

IOP Conf. Series: Journal of Physics: Conf. Series **1014** (2018) 012004 doi:10.1088/1742-6596/1014/1/012004

model (see ref. [17]). The damping of the shell effects decreases the possibility of the deformed mononucleus of reaching the CN equilibrium shape, and the mononucleus breaks down into two fragments without to reach the CN stage. Therefore, the fast fission cross section $\sigma_{\rm ff}$ is calculated by summing the contributions of the partial fusion cross sections with ℓ values corresponding to the range $\ell_f < \ell < \ell_d$ leading to the formation of the mononucleus,

$$\sigma_{\rm ff}(E_{\rm c.m.}) = \sum_{\ell=\ell_f}^{\ell_d}(2\ell+1)\sigma_{\rm cap}(E_{\rm c.m.},\ell)P_{\rm CN}(E_{\rm c.m.},\ell). \quad (4)$$

As the advantage of our modular system of nuclear reaction codes we stress the possibility to include into calculation explicitly the effect of the orbital angular momentum on the capture, fusion and survival probability. This fact allows us to analyze the role of the entrance channel effects on the evaporation residue cross section [7].

2. Fusion probability $P_{\rm CN}$ determinations

As discussed in Introduction by regarding the use of formula (1) for the determination of $P_{\rm CN}$ values of many kinds of nuclear reaction, the reason for the relevant discrepancy between the measured data and theoretical values of $P_{\rm CN}$ is connected with the experimental difficulties in the identification of the fusion-fission fragments produced by fission of the compound nucleus to determine the fusion cross section. The mass distribution of the fast fission and sometimes quasifission processes overlaps with the mass symmetrical fusion-fission product distributions. The ability of the correct extraction of the fusion-fission cross section from the mixed data of the reaction products decreases due to intensive population of the mass-symmetric region by the fast fission or/and the quasifission fragments by the increase of $E^*_{\rm CN}$. Therefore, the extraction of the fusion cross section from the experimental data is strongly affected by the underestimation of contribution of the fast fission and quasifission fragments. This problem is inherent to all kinds of reactions, leading to the formation of superheavy nuclei, when experimentalists selecting only mass symmetric fragments with the mass numbers in the range $A_{\rm CN}/2\pm20$, assume such products belong only to the fusion-fission process. In fact, this assumption is completely doubtful because the yield of mass symmetric fragments produced by the quasifission and fast fission processes are competitive and often some orders of magnitude higher than the ones produced by the fusion-fission process (see for example, refs. [12, 13].

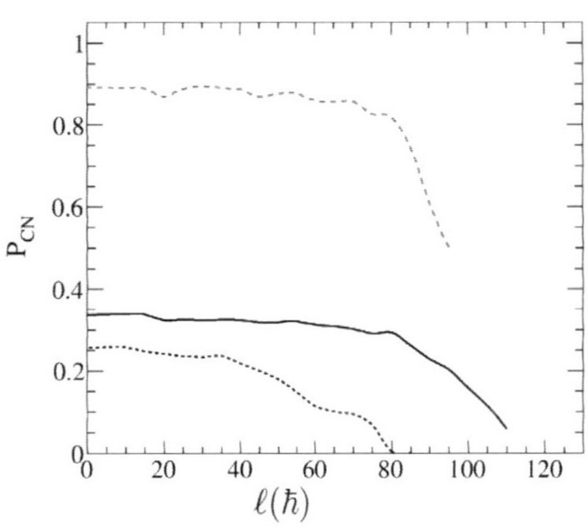

Figure 1. (Color on-line) The $P_{\rm CN}$ fusion probability as a function of the angular momentum ℓ for the ^{16}O+^{186}W very asymmetric reaction at excitation energy $E^*_{\rm CN}$=75 MeV (dashed line) of the ^{202}Pb CN is shown; in the same figure, the $P_{\rm CN}$ values vs ℓ for the ^{48}Ca+^{154}Sm less asymmetric reaction at excitation energies $E^*_{\rm CN}$= 49 and 63 MeV, respectively (dotted and full lines), are also reported.

Moreover, in order to show the sensitivity of the $P_{\rm CN}$ with the angular momentum ℓ at two different excitation energies of 49 and 63 MeV (dotted and full lines, respectively) of the formed ^{202}Pb compound nucleus, we present in Fig. 1 our results obtained for the ^{48}Ca+^{154}Sm less asymmetric reaction. We also present in the same figure the $P_{\rm CN}$ values vs ℓ obtained for the ^{16}O+^{186}W very asymmetric reaction, leading to the same ^{202}Pb CN at $E^*_{\rm CN}$=63 MeV (dashed line). This figure clearly shows the strong dependence of the $P_{\rm CN}$ results on the excitation energy $E^*_{\rm CN}$ and/or of the beam-target combination in the entrance channel.

Therefore, the comparison of the $P_{\rm CN}$ values for different conditions of reactions can be made and understood only if somebody is able to explain the reasons for different results and trend due to the entrance channel effects and/or characteristics of the reaction mechanism (for details see [11]).

3. Survival probability $W_{\rm sur}$ determinations and ER residual nuclei formation

Another important problem in the analysis of the experimental data is related with the determination of the reliable $B_{\rm fis}$ fission barrier values of excited nuclei reached along the deexcitation cascade of CN in order to correctly estimate the overall $W_{\rm sur}(E^*_{\rm CN})$ survival probability against fission which is used to calculate the total ER cross section vs $E^*_{\rm CN}$. Indeed, the experimental determinations of the ERs values may be uncertain whether any of α-decay lifetime of reached nuclei along the deexcitation cascade is less than few μs. The uncertainty in determination of the ERs values appears even more strongly when the process of the deexcitation of CN by the evaporation of the charged particles is neglected in comparison with the evaporation of neutral particles. In fact, in many heavy-ion reactions leading to formation of the heavy and superheavy compound nuclei, the total evaporation residue cross sections (when the charged particles are taken into account too) are much higher than the ones obtained for the evaporation of neutrons only. For example, the $\sigma_{\rm ERtot}/\sigma_{\rm ER-xn}$ ratio is 5–8 times for the ^{26}Mg+^{248}Cm and ^{36}S+^{238}U reactions (leading to the same ^{274}108 CN called ^{274}Hs*) and it can reach even 1 or 2 orders of magnitude for some other reactions (see for example [7,18]). Therefore, neglecting the contribution of the charged particles in the determination of evaporation residue nuclei without the possibility of knowing the effect on the final results is doubtful. We can conclude here that the complete evaporation residue cross section $\sigma_{\rm ERtot}$ can be determined correctly if there is a possibility of full detection of all the total evaporation residue nuclei in the reaction.

The evaporation residue cross sections at the given values of the CN excitation energy E^*_x at each step x of the deexcitation cascade by the advanced statistical model [9]

$$\sigma^{(x)}_{\rm ER}(E^*_x) = \Sigma^{\ell_d}_{\ell=0}(2\ell+1)\sigma^{(x)}_{\rm ER}(E^*_x,\ell), \tag{5}$$

where $\sigma^{(x)}_{\rm ER}(E^*_x,\ell)$ is the partial cross section of ER formation obtained after the emission of particles $\nu(x)$n + $y(x)$p + $k(x)\alpha$ + $s(x)$ (where $\nu(x)$, y, k, and s are numbers of neutrons, protons, α-particles, and γ-quanta) from the intermediate nucleus with excitation energy E^*_x at each step x of the deexcitation cascade by the formula (for more details, see papers [2,9,11,14]):

$$\sigma^{(x)}_{\rm ER}(E^*_x,\ell) = \sigma^{(x-1)}_{\rm ER}(E^*_{x-1},\ell)W^{(x)}_{\rm sur}(E^*_{x-1},\ell). \tag{6}$$

In calculation of the $W^{(x)}_{\rm sur}(E^*_{x-1},\ell)$ the fission barrier is used as a sum of the parametrized macroscopic fission barrier $B^m_{fis}(\ell)$ depending on the angular momentum ℓ [16] and the microscopic (shell) correction $\delta W = \delta W_{sad} - \delta W_{gs}$ due to shell effects; by considering the large deformation of a fissioning nucleus at the saddle point, δW_{sad} is much smaller than the δW_{gs} value and the microscopic shell correction δW to the fission barrier can be expressed by the relation $\delta W \cong -\delta W_{\rm gs}$. Therefore, an effective fission barrier, as a function of ℓ and T for each excited nucleus formed at various steps along the deexcitation cascade of CN, is calculated by the expression

$$B_{\rm fis}(\ell,T) = c\ B^m_{fis}(\ell) - h(T)\ q(\ell)\ \delta W, \tag{7}$$

where the factor c was set to 1 in all our calculations and $h(T)$ and $q(\ell)$ represent the damping functions of the nuclear shell correction (usually it is $\delta W < 0$) with the increase of the excitation energy E^* and ℓ angular momentum, respectively [9]:

$$h(T) = \{1 + \exp[(T - T_0)/d]\}^{-1} \tag{8}$$

and

$$q(\ell) = \{1 + \exp[(\ell - \ell_{1/2})/\Delta\ell]\}^{-1}, \tag{9}$$

where, in Eq. (8), $T = \sqrt{E^*/a}$ represents the nuclear temperature depending on the excitation energy E^* and the level density parameter a, $d = 0.3$ MeV is the rate of washing out the shell corrections with the temperature, and $T_0 = 1.16$ MeV is the value at which the damping factor $h(T)$ is reduced by 1/2. Analogously, in Eq. (9), $\Delta\ell = 3\hbar$ is the rate of washing out the shell corrections with the angular momentum, and $\ell_{1/2} = 20\hbar$ is the value at which the damping factor $q(\ell)$ is reduced by 1/2.

By regarding the determination of intrinsic level density ρ_{int} which takes into account the density of the intrinsic excitations, the nucleus is considered as a system made up to non-interacting Fermi gases (proton gas and neutron gas) at the same thermodynamic temperature T. It is supposed that each of the two gases is in thermodynamic equilibrium and that the excitation energy E^* is distributed in a statistical way between two gases. In this context for the intrinsic level density parameter a we use the general expression [19] especially tailored to account for the shell effects in the level density

$$a(E^*) = \tilde{a}\left\{1 + \delta W\left[\frac{1 - exp(-\gamma E^*)}{E^*}\right]\right\} \tag{10}$$

where $\tilde{a} = 0.094 \times A$ is the asymptotic value that takes into account the dependence on the mass number A, and γ =0.0064 MeV^{-1} is the parameter which accounts for the rate at which shell effects wash out with excitation energy for neutron or other light particle emission. The general expression (10) works well also for deformed prolate or oblate nuclei. Physically, the disappearance of the shell effects with E^* excitation energy may be seen as a rearrangement of the shell-model orbitals in such a way that the shell gap between orbitals close to the Fermi energy vanishes. The value of the γ parameter was obtained [19] by fitting the observed density of neutron resonances.

In order to determine the a_{fis} level density parameter in the fission channel we use the relation $a_{fis}(E^*) = a_n(E^*) \times r(E^*)$ found in [20] where $r(E^*)$ is given by the relation

$$r(E^*) = \frac{[exp(-\gamma_{fis}E^*) - (1 + \frac{E^*}{\delta W})]}{[exp(-\gamma E^*) - (1 + \frac{E^*}{\delta W})]} \tag{11}$$

with $\gamma_{fis} = 0.024$ MeV^{-1}.

To calculate the intrinsic level density $\rho_{int}(E^*, \ell)$ we use the general expression

$$\rho_{\text{int}}(E, J) = \frac{1}{16\sqrt{6\pi}}\left[\frac{\hbar^2}{\mathcal{J}_{\|}}\right]^{1/2} a^{-1/4} \times \sum_{k=-J}^{J} [E - E_{\text{rot}}(k)]^{-5/4} e^{2\{a[E - E_{\text{rot}}(k)]\}^{1/2}} \tag{12}$$

The collective level density ρ_{coll} calculated in the adiabatic approach, valid at low excitation energies, takes into account in addition to the intrinsic excitations also the rotational and vibrational excitation states by the collective enhancement factor $K_{\text{coll}}(E^*)$:

$$\rho_{\text{coll}}^{\text{adiab}}(E^*, J) = \rho_{\text{int}}(E^*, J) \times K_{\text{coll}}^{\text{adiab}}(E^*)$$

where $K_{\text{coll}}^{\text{adiab}}(E^*)$ is given by the simple multiplication of the two $K_{\text{rot}}^{\text{adiab}}(E^*)$ and $K_{\text{vibr}}^{\text{adiab}}(E^*)$ enhancement factors; therefore, $\rho_{\text{coll}}^{\text{adiab}}(E^*, J)$ is determined (for more details see Appendix B) as

$$\rho_{\text{coll}}^{\text{adiab}}(E^*, J) = \rho_{\text{int}}(E^*, J) \times K_{\text{rot}}^{\text{adiab}}(E^*) \times K_{\text{vibr}}^{\text{adiab}}(E^*). \tag{13}$$

In the cited paper [11] we give many other details regarding the intrinsic ρ_{int} and collective ρ_{coll} level density determinations, the fission Γ_{fis} and particle-x Γ_x decay widths, and we show the sensitivity of the model on final reaction products by using mass asymmetric and almost symmetric reactants in the entrance channel.

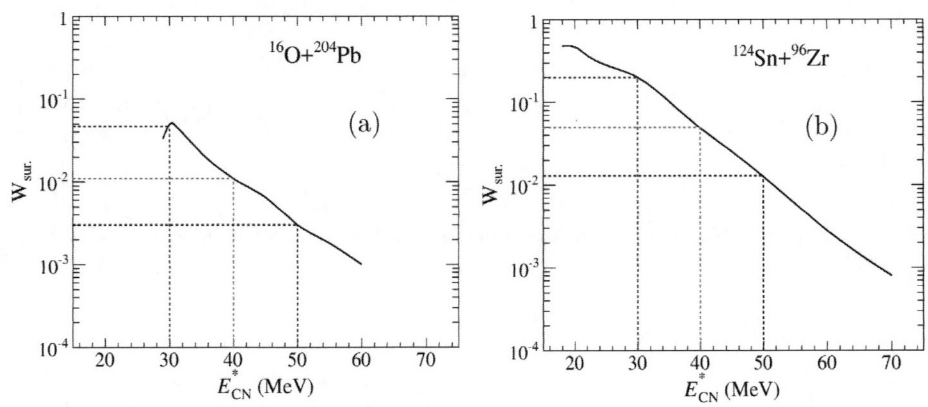

Figure 2. (Color on-line) The W_{sur} survival probability against fission along the deexcitation cascade for the ^{16}O+^{204}Pb reaction (a) and the ^{124}Sn+^{96}Zr reaction (b), both leading to the ^{220}Th CN, as a function of the E^*_{CN} excitation energy.

By comparing the W_{sur} values obtained for two very different entrance channels leading to the formation of the ^{220}Th CN at excitation energies E^*_{CN}=30, 40 and 50 MeV, it is possible to observe the different effects of the angular momentum distribution on the surviving probability excitation functions. The W_{sur} values for the CN formed in the almost symmetric ^{124}Sn+^{96}Zr reaction are greater than the W_{sur} values corresponding to the one in the very asymmetric ^{16}O+^{204}Pb reaction by factors of 3.5, 4.2, and 4.5 times at the E^*_{CN}=30, 40 and 50 MeV, respectively. Certainly the increase of the beam energy $E^*_{\text{c.m.}}$ leads to an increase of the excitation energy E^*_{CN} of CN and to an extension of the angular momentum distribution of the CN formed in these reactions. But the behavior of the extension of the angular momentum distribution is different for the two reactions. The ℓ angular momentum range is larger for the ^{220}Th CN obtained in the very asymmetric ^{16}O+^{204}Pb reaction than for the CN obtained in the almost symmetric ^{124}Sn+^{96}Zr reaction at the same considered E^*_{CN} excitation energy. This difference in the angular momentum ℓ interval increases with the increase of E^*_{CN} since the size of the potential well in the nucleus-nucleus interaction for the almost symmetric ^{124}Sn+^{96}Zr reaction is smaller than the one for the very asymmetric reaction like ^{16}O+^{204}Pb. Therefore, the number of the angular momentum ℓ contributing to fusion in the ^{124}Sn+^{96}Zr reaction is smaller. Moreover, the fusion probability P_{CN} strongly decreases by increasing the angular momentum due to the increase of the B^*_{fus} intrinsic fusion barrier and due to a decrease of the B_{qf} quasifission barrier for the symmetric reaction (see for example ref. [21]). The quasifission barrier is the depth of the potential well in the nucleus-nucleus interaction.

Obviously, the fusion cross section at a considered $E^*_{\rm CN}$ value of CN for the ^{124}Sn+^{96}Zr symmetric reaction is smaller than the one of the ^{16}O+^{204}Pb very asymmetric reaction, but with a smaller ℓ interval of the formed CN by the symmetric reaction has a greater W_{sur} survival probability to fission in comparison with the very asymmetric reaction.

Moreover, as discussed in our paper [18], it is difficult to estimate the $\sigma_{\rm ERtot}$ cross section in experiment, when the emission of charged particles present also, because not all of residue nuclei can be identified. Therefore, apart from the uncertainties which are inherent to the theoretical predictions, there are ambiguities in the estimation of the fusion cross sections by the analysis of experimental data.

In our code the fission and particle decay widths $\Gamma_{\rm fis}$ and $\Gamma_{\rm x}$ are calculated by the formulas

$$\Gamma_{\rm fis}(E,J) = \frac{1}{2\pi\rho(E,J)} \int_0^{E-E_{\rm sad}(J)} \rho_{\rm fis}(E-E_{\rm sad}(J)-\epsilon, J) \times T_{\rm fis}(E-E_{\rm sad}(J)-\epsilon)d\epsilon, \tag{14}$$

and

$$\Gamma_x(E,J) = \frac{1}{2\pi\rho(E,J)} \sum_{J'=0}^{\infty} \sum_{j=|J'-J|}^{J'+J} \times \int_0^{E-B_x} \rho_{\rm x}(E-E_{\rm x}-\epsilon, J')T_{\rm x}^{\ell,j}(\epsilon)d\epsilon, \tag{15}$$

where the subscript fis and x refer, respectively, to the fission process and particle-x emission channels (neutron, proton, α, and γ), and primes are used to mark an intermediate excited nucleus after particle emission and $E_{\rm sad}(J)$ is the energy of the decaying nucleus at the saddle point with angular momentum ℓ and total spin J.

Formulas (14) and (15) ρ, $\rho_{\rm fis}$ and $\rho_{\rm x}$ represent the collective level densities of the formed excited nucleus, the level density of the excited nucleus at the saddle point configuration for orbital angular momentum ℓ, and the level density of the reached subsequent excited nucleus after particle-x emission for orbital angular momentum ℓ, respectively. $T_{\rm x}^{\ell,j}$ is an optical-model transmission coefficient for particle-x with angular momentum ℓ coupled with particle spin to give j, and the fission transmission coefficient $T_{\rm fis}$ in the Hill-Wheeler approximation is given by $T_{\rm fis} = \{1 + exp[-2\pi(E^* - E_{\rm sad}(J) - \epsilon)/\hbar\omega]\}^{-1}$, with $\hbar\omega = 1$ MeV.

It is useful to observe that in formula (14) the calculation of $\Gamma_{\rm fis}$ fission width is characterized in the nominator by the level density $\rho_{\rm fis}$ of the excited nucleus that reaches the saddle point state with angular momentum ℓ and total spin J - weighted by the transmission coefficient $T_{\rm fis}$ - and in the denominator by the level density $\rho(E,J)$ of the same excited and rotating nucleus at the statistical equilibrated state with energy E and spin J, while in (15) the calculation of the $\Gamma_{\rm x}$ width for particle-x emission - weighted by the $T_{\rm x}$ transmission coefficient - is characterized in the nominator by the level density $\rho_{\rm x}(E - B_{\rm x} - \epsilon, J')$ of the intermediate excited nucleus.

In order to calculate the collective level density in the non-adiabatic approach $\rho_{\rm coll}^{\rm non-adiab}(E^*, J)$ - necessary when the CN is formed at higher $E^*_{\rm CN}$ excitation energies - we used a damped collective enhancement function $q(E^*, \beta)$ with the aim to account the coupling of the collective to intrinsic degrees of freedom due to the nuclear viscosity because while it is acceptable to treat collective modes within the adiabatic approximation at low energies, it is rather unlikely that at high energies the adiabatic assumption still holds, due to the coupling of the elementary modes to the collective ones. So we introduce a certain general function for damping collective effects in the level density, depending strongly on the excitation energy and deformation of the nucleus. This is expressed in a decrease of the collective enhancement coefficient $K_{\text{coll}}(E^*)$ when the excitation energy increases; therefore, the following expression was assumed [22, 23]

$$K_{\text{coll}}^{\text{non-adiab}}(E^*) = \left\{[K_{\text{rot}}^{\text{adiab}}(E^*) - 1]q(E^*, \beta) + 1\right\} \times \left\{[K_{\text{vibr}}^{\text{adiab}}(E^*) - 1]q(E^*, \beta) + 1\right\} \quad (16)$$

where

$$q(E^*, \beta) = exp[-E^*/E_1(\beta)]. \quad (17)$$

The expression of damping function is used in the fission and neutron (or other light particles) emission channels in the same way [24]. In formula (17) the following expression for $E_1(\beta)$ was assumed

$$E_1(\beta) \cong 170 \times A^{\frac{1}{3}}\beta^2 \quad (18)$$

in order to reach the better agreement between calculated values of fission cross sections and experimental ones for a wide set of nuclear reactions leading to compound nuclei lighter than lead, preactinide and actinide compound nuclei, and also for nuclei with Z>100. At the same time, the comparison between theoretical estimation of evaporation residue cross sections and experimental determinations have contributed to the choice of the damping function q(E*,β) to the K_{coll} collective enhancement coefficient given in formula (17) together with the expression (18).The consequence of the quadratic dependence of $E_1(\beta)$ on β (see relation (18)) is that the damping at the saddle point ($\beta \simeq 0.6 - 0.8$) is negligible in the high E^* excitation energy region, while the deviations of q(E^*,β) from unity are already considerable for the neutron channel ($\beta \simeq 0.2 - 0.3$) at low E^* excitation energy values.

4. Analysis of ER excitation functions for the ^{82}Se+^{138}Ba reaction

We analyze and present the ER excitation functions for the ^{82}Se+^{138}Ba almost mass symmetric ($\eta = 0.255$) reaction since it is possible to explore the low excitation energy range starting from the value E^*_{CN} =12 MeV that corresponds to the threshold energy for the ^{220}Th CN formation. In comparison with the fusion excitation functions obtained for the mass asymmetric ^{16}O+^{204}Pb ($\eta = 0.855$) and ^{40}Ar+^{180}Hf ($\eta = 0.636$) reactions, the maximum value of the fusion excitation function for the reaction induced by the ^{82}Se beam is about two orders of mangitude lower than the value obtained for the reaction induced by the ^{16}O beam, and about one order of magnitude lower than the value obtained for the reaction induced by the ^{40}Ar beam (see Fig.6 in [7]). The strong difference in the fusion excitation functions is related to the size of the potential well in the nucleus-nucleus interaction at the capture stage. The increase of the beam energy from near the Coulomb barrier energies leads to increase of number of partial waves contributing to capture and complete fusion. The Coulomb repulsion is stronger for the almost symmetric reactions in comparison with the one for the asymmetric reactions. Therefore, strong repulsion forces make the potential well shallower, reducing consequently the maximum number of the partial waves [ℓ_d(E)] leading to capture; consequently, the capture cross section decreases. This effect is related to the dependence of the fusion barrier B_{fus} and quasifission barrier B_{qf} on the angular momentum values.

Fig.3 shows in panels (a), (b), (c) and (d) the experimental values obtained in ref. [25] for the individual $\sigma_{\text{ER}_{\text{xn}}}$ cross sections after 3n, 4n, 5n, and 6n+7n neutron emission, respectively. In order to compare our calculated results for the above individual $\sigma_{\text{ER}_{\text{xn}}}$ excitation functions vs

IOP Conf. Series: Journal of Physics: Conf. Series **1014** (2018) 012004 doi:10.1088/1742-6596/1014/1/012004

E^*_{CN} with the avilable measurements, we also report in the above panels dashed lines representing the values obtained by our code. The results of the theoretical calculations are in good agreement with experimental measurements for all panels.

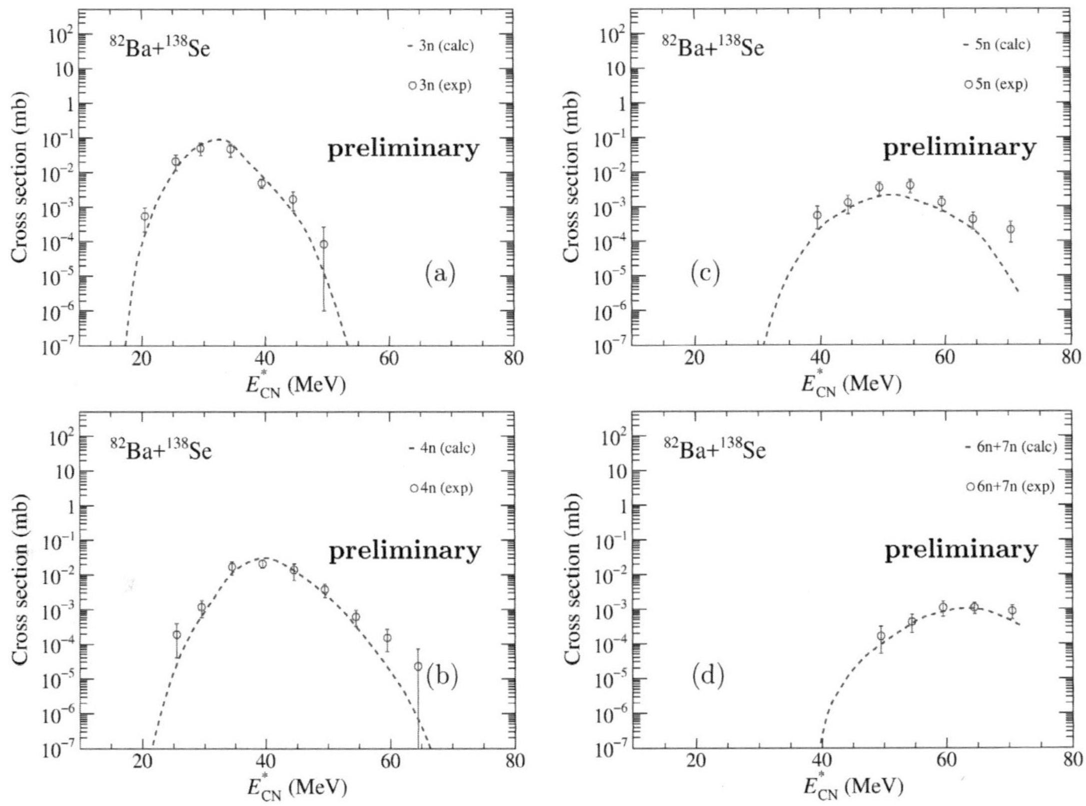

Figure 3. (Color on-line) The individual ER cross sections after 3n (panel (a)), 4n (b), 5n (c), and 6n+7n (d) neutron emission vs. E^*_{CN} for the ^{82}Se+^{138}Ba reaction leading to the ^{220}Th CN. The points represent the experimental values taken from ref. [25]. In each panel, the dashed line represents the related theoretical excitation function obtained by our code.

Fig.4 shows in panels (a), (b), (c) and (d) the experimental values given in ref. [25] for the individual ER cross sections after α1n+1n, α2n+2n, α4n+α5n and α6n+α7n emissions, respectively. Also in this case we report in the above panels the dashed lines representing the theoretical values obtained by our code. The related results of calculation for the indicated individual channels are in good agreement with the experimental measurements in all panels.

Moreover, Fig.5 shows in panels (a), (b), (c), and (d) the experimental values given in ref. [25] for the individual ER cross sections after p3n, p4n, p5n, and p6n+p7n emissions, respectively. Analogously, also in this case, we report in the above panels by dashed lines the theoretical results obtained by our code; we register still a good agreement of these results with the experimental ones given in the above-mentioned panels.

The comparison made on the results presented in Figs. 3, 4, and 5 ensures the goodness of the experimental determinations and the reliability of the results obtained by our theoretical procedures, but unfortunately the complete final result is that the experimental determinations cover only a partial set of the possible charged particle emissions in the investigated ^{82}Se+^{138}Ba reaction. In fact in our theoretical calculations we have extended the determination of other

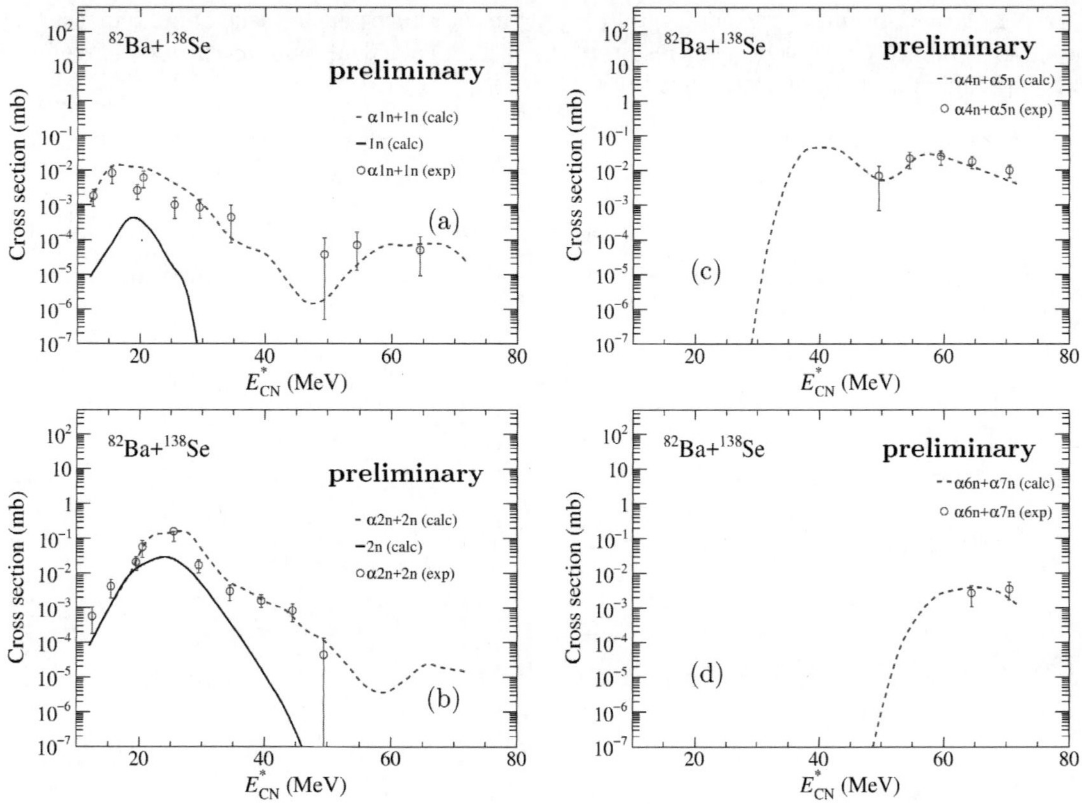

Figure 4. (Color on-line) As Fig.3 but for the α1n+1n (panel (a)), α2n+2n (b), α4n+α5n (c), and α6n+α7n (d) channels.

relevant contributions of charged particle emissions as for example the αpxn, 2αxn, 2αpxn, 3αxn, and 3αpxn channels leading to the formation of other various ER evaporation residue nuclei.

At last, in Fig. 6 we report by dashed line the total contribution of ER cross sections after neutron emission only, by thin full line the contribution of ER cross section when also the αxn, and pxn channels (considered in ref. [25]) are included, and by thick full line when also the other αpxn, 2αxn, 2αpxn, 3αxn, and 3αpxn channels are considered. Therefore, this last line represents the true total ER excitation function vs $E^*_{\rm CN}$ obtained in the ^{82}Se+138Ba reaction. Moreover, in the insert of Fig. 6 we report the $W_{\rm sur}$ survival probability excitation function vs $E^*_{\rm CN}$ for the ^{220}Th CN formed by the considered reaction in the entrance channel. At $E^*_{\rm CN}$ >40 MeV $W_{\rm sur}$ decreases by 2-3 orders of magnitude to cause of the dominant fission process. From this figure it is possible to observe that in the about $E^*_{\rm CN}$ 27-30 MeV of excitation energy range the ratio between the yield of the total ER cross section (contributed by evaporation of all possible charged and neutral particles) and the yield of the ER cross section obtained for evaporation of neutrons only, is about a factor 20 confirming the decisive role of emission of charged particles along the deexcitation cascade of CN in order to form the residual nuclei. Moreover, it is possible to observe in the same figure that the true total ER cross section obtained for possible contributions of evaporations of charged and neutron particles (represented by the thick full line) is at least a factor 2-4 higher than the sum of the ER cross sections given in ref. [25] in the 27-70 MeV $E^*_{\rm CN}$ energy range.

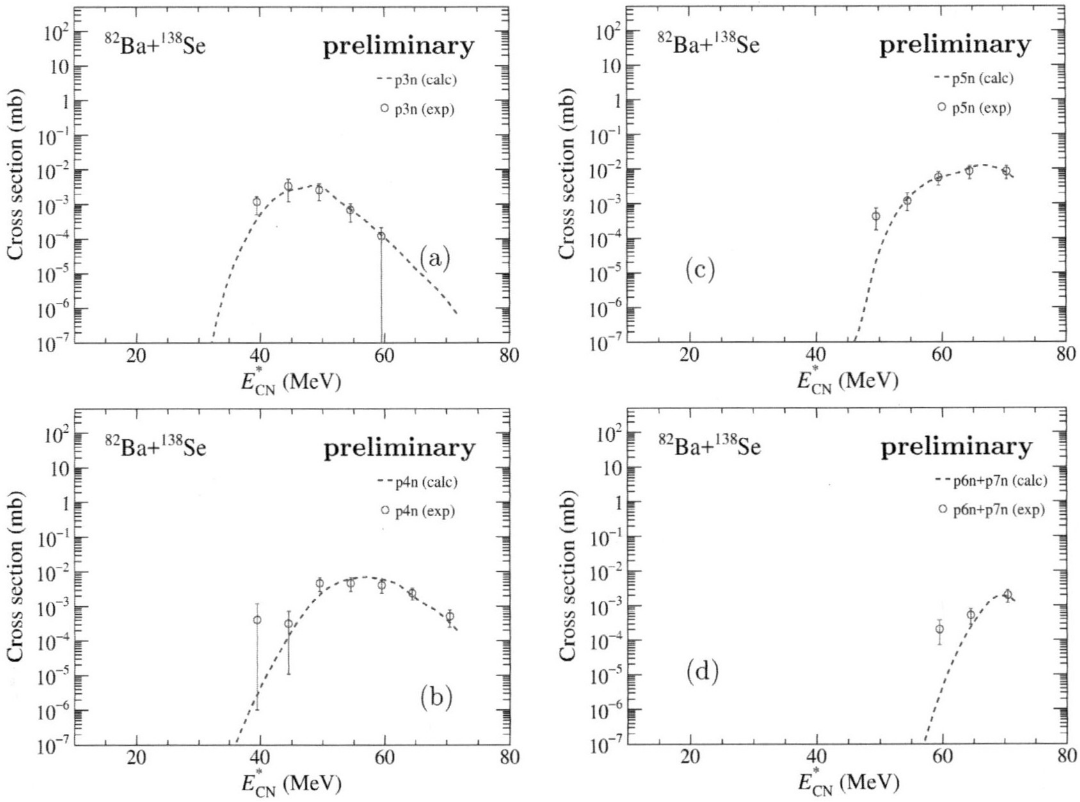

Figure 5. (Color on-line) As Fig.3 but for the p3n (panel (a)), p4n (b), p5n (c), and p6n+pn (d) channels.

Therefore this study demonstrates that many times, because of the instrinsic limitations of the experimental apparatus to identify all real contributions [25], or simplified assumptions are made in the analysis of experimental results, or when one adopts calculation procedures based on parameter-free phenomenological models, the obtained ER results are inconsistent and unreliable.

Moreover, in order to demonstrate the sensitivity of the method and the relevant effect of the entrance channel on the fusion probability $P_{\rm CN}$ values for two very close reactions, we present in Fig. 7 the results for the ^{26}Mg+^{248}Cm and ^{25}Mg+^{248}Cm reactions leading to the ^{274}Hs CN and ^{273}Hs CN, respectively, different only for one neutron.

Fig. 7 (a) shows the capture, fusion and quasifission excitation functions for the reaction induced by the prolate deformed (0.363) ^{25}Mg beam; panel (b) shows the same mentioned excitation functions for the reaction induced by the deformed oblate (-0.309) ^{26}Mg beam.

To a simple visual recognition of the lines reported in panels (a) and (b) show the presence of a small difference of 2 MeV in the threshold energy value and some quite similar appreciable difference in the trend of quasifission, but from Fig. 8 (that shows the fusion probability excitation function $P_{\rm CN}$) appears instead to have a complete different trend of the $P_{\rm CN}$ excitation function for the two mentioned reactions vs the collision energy $E_{\rm c.m.}$-$E_{\rm B}$ as compared with the Bass barrier.

This example demonstrates the importance of the type of deformation (oblate for the ^{26}Mg beam, prolate for the ^{25}Mg beam) for a nucleus (beam or target) in the entrance channel and

IOP Conf. Series: Journal of Physics: Conf. Series **1014** (2018) 012004 doi:10.1088/1742-6596/1014/1/012004

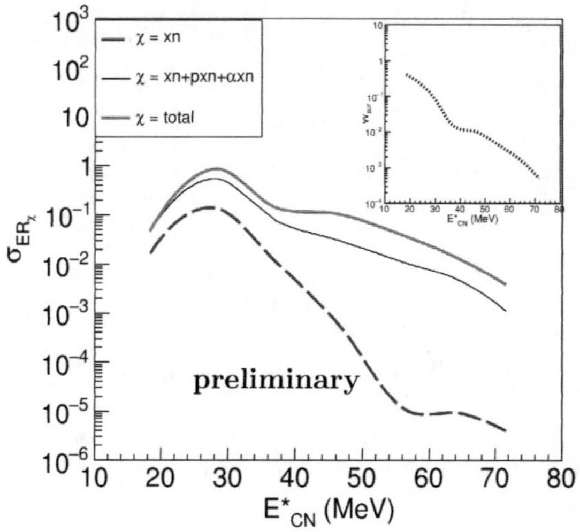

Figure 6. (Color on-line) The ER excitation functions vs $E^*_{\rm CN}$ calculated by our code for the ^{82}Se+^{138}Ba reaction leading to the ^{220}Th CN: dashed line represents the ER contribution after neutron emission only considered in Fig.3; thin full line represents the sum of ER contributions after xn, pxn, and αxn particle emissions considered in Fig.3, 4, and 5; thick full line represents the sum of ER contributions considered in Figs.3, 4, 5, with in addittion the above-mentioned 2αxn, 2αpxn, 3αxn, and 3αpxn contributions. In the insert, the dotted line represents the excitation function vs $E^*_{\rm CN}$ of the real $W_{\rm sur}$ survival probability calculated by our code when the total ER contributions represented by the thick full line is considered.

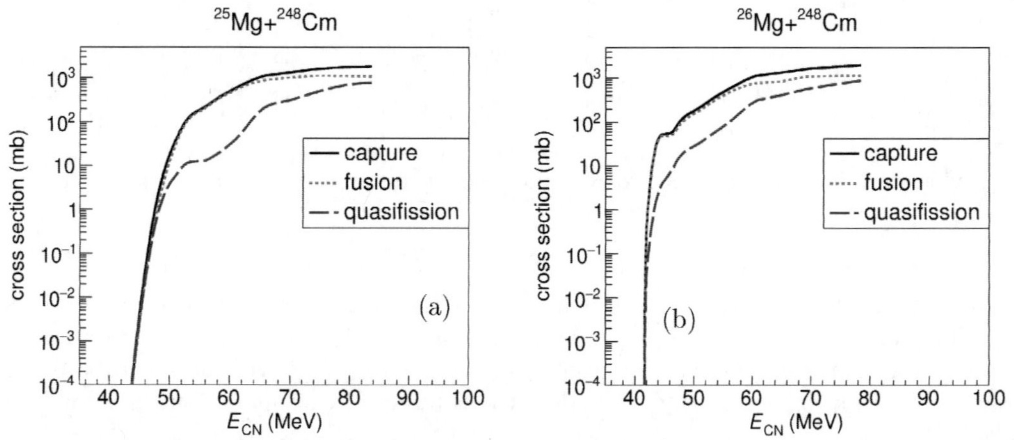

Figure 7. (Color on-line) Comparison of the capture, fusion, and quasifission excitation functions calculated for: (a) the ^{25}Mg+^{248}Cm reaction leading to the ^{273}Hs CN; (b) the ^{26}Mg+^{248}Cm reaction leading to the ^{274}Hs CN.

how this strongly influences the reaction mechanism and the determination of the reaction final products. The present specific case (by comparing some results of the ^{25,26}Mg+^{248}Cm reactions leading to the ^{273,274}Hs compound nuclei, respectively) was already presented in our paper [11] and we discussed about the unreliable result found in paper [26] on the value $\Gamma_{\rm n}/\Gamma_{\rm total}$ =0.89±0.13 for the probability of neutron emission at first step of deexcitation of the ^{274}Hs CN with $E^*_{\rm CN}$ =63 MeV of excitation energy (and consequently a value of $\Gamma_{\rm fis}/\Gamma_{\rm total} \simeq$11 for the fission probability) by using the formula (7) of ref. [26] $(\Gamma_{\rm n}/\Gamma_{\rm total})_1 = \nu^{274}_{\rm pre}/(1+\nu^{273}_{\rm pre})$. But, in [26] the authors affirmed that "For the range of the $B_{\rm fis}$ values predicted by current

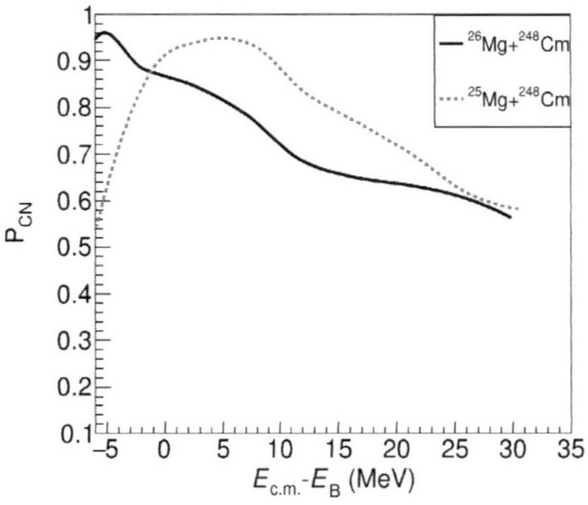

Figure 8. (Color on-line) $P_{\rm CN}$ fusion probability vs $E_{\rm c.m.} - E_{\rm B}$ for: (a) the ^{25}Mg + ^{248}Cm reaction leading to ^{273}Hs CN (dashed line); (b) ^{26}Mg + ^{248}Cm reaction leading to ^{274}Hs CN (full line).

models, there is very little sensitivity of $B_{\rm fis}$. To get values of $\Gamma_{\rm n}/\Gamma_{\rm total}$ similar to the measured value, one must invoke values of $B_{\rm fis}$ ~11 MeV, a number not consistent with current theoretical models for superheavy nuclei." Moreover, the above-mentioned formula (7) of ref. [26] was obtained by assuming that the fission-neutron emission competition of the reaction induced by the ^{26}Mg beam on the ^{248}Cm target after 1 neutron emission from the ^{274}Hs CN with excitation energy $E^*_{\rm CN}$=63 MeV reaches the intermediate excited nucleus ^{273}Hs with excitation energy of 53 MeV, it is similar in all the characteristics to the ^{273}Hs CN obtained in the reaction induced by the ^{25}Mg beam of appropriate energy on the same ^{248}Cm target in order to reach the ^{273}Hs CN with the same excitation energy of $E^*_{\rm CN}$ =53 MeV. By realizing this condition for the two above-mentioned reactions induced by the ^{26}Mg and ^{25}Mg beams on the ^{248}Cm target, the authors [26] affirmed that the following deexcitation cascade from the ^{273}Hs intermediate excited nucleus with the excitation energy of 53 MeV formed by the ^{26}Mg + ^{248}Cm reaction after 1 neutron emission from the ^{274}Hs CN at $E^*_{\rm CN}$ = 63 MeV is completely the same for the $\nu^{273}_{\rm pre}$ and $(\Gamma_{\rm n}/\Gamma_{\rm total})_{273}$ values of the deexcitation cascade from the ^{273}Hs CN at $E^*_{\rm CN}$ =53 MeV formed by the ^{25}Mg + ^{248}Cm reaction.

Instead, in [11] was explained that this assumption in data analysis [26] does not hold, for two reasons: **a)**in the ^{25}Mg+^{248}Cm reaction, the ^{273}Hs CN formation has a certain excitation energy $E^*_{\rm CN}$ (for example $E^*_{\rm CN}$=53 MeV) with respect to a certain ^{25}Mg energy beam; the energy spread (of some hundred keV) of $E^*_{\rm CN}$ for the ^{273}Hs CN is only related to the determination of the energy beam and also to the thickness of the target. Analogously, in the ^{26}Mg+^{248}Cm reaction, the ^{274}Hs CN formation has a certain excitation energy $E^*_{\rm CN}$ (for example $E^*_{\rm CN}$=63 MeV) with respect to a certain ^{26}Mg energy beam and related energy resolution, but the formation of the intermediate excited ^{273}Hs nucleus, reached after 1 neutron emission from the ^{274}Hs CN, is characterized by a broader excitation energy peak (of some MeV) with respect to the kinetic energy distribution of the neutron emitted from the ^{274}Hs CN. **b)** In the present specific case of the ^{25}Mg and ^{26}Mg induced reactions, the dynamic deformation of the ^{25}Mg and ^{26}Mg beams are very different at collision with the ^{248}Cm target, because the ^{25}Mg nucleus has a prolate static deformation (+0.363) while ^{26}Mg has a oblate static deformation (-0.309). These different conditions of the entrance channel affect the dynamic deformation of the ^{274}Hs CN and ^{273}Hs CN formed by the two ^{26}Mg (^{25}Mg)+^{248}Cm reactions, and the intermediate excited nucleus ^{273}Hs (for example at E^*=53 MeV of excitation energy) reached after 1 neutron emission from the ^{274}Hs CN is dynamically different with respect to the ^{273}Hs CN (at $E^*_{\rm CN}$=53 MeV) formed by the ^{25}Mg+^{248}Cm reaction. Moreover, the result we obtained for the probability of neutron emission $\Gamma_{\rm n}/\Gamma_{\rm total}$ =0.17 for the ^{274}Hs CN at $E^*_{\rm CN}$ =63 MeV is completely different from the value $\Gamma_n/\Gamma_{total} = 0.89 \pm 0.13$ deduced in experiment [26]. Instead, our result (Γ_n/Γ_{total} =0.17) allows us to affirm that the fission probability $\Gamma_{fis}/\Gamma_{total}$ =0.83 at first step of deexcitation of the ^{274}Hs CN at E^*_{CN} = 63 MeV is the dominant process to cause of the small effective fission barrier B_{fis}=0.72 MeV contributed only by the shell correction value that is damped by the nuclear temperature T (namely the related excitation energy E^*) and the angular momentum ℓ, since the contribution to the macroscopic rotating liquid drop model is zero. Following this procedure we calculate for the ER cross section about 0.6 pb at E^*_{CN} =63 MeV of the ^{274}Hs CN by the ^{26}Mg+^{248}Cm reaction.

This last result is consistent with the results obtained in ref. [27], while by assuming the value Γ_n/Γ_{total} =0.89 given in [11] we estimate to reach a ER value of about 10^{-3}mb that is at least 6 orders of magnitude higher than the experimental determination [27]. Moreover, the determinations of ν^{274}_{pre} and ν^{273}_{pre} values and the use of formula (7) of ref. [26] is not a reliable way to determine the $(\Gamma_n/\Gamma_{total})_1$ value since the method and procedures do not take into account the detailed dynamic effects of reactions. In fact, if for example one uses the values of ν^{274}_{pre}=1.7, 2, 2.7, 3, and 4 at $E^*(^{274}$Hs) = 63 MeV with the corresponding of ν^{273}_{pre}=1.37, 1.61, 2.17, 2.42, and 3.22 at $E^*(^{273}$Hs) = 53 MeV that assure the same $\nu^{274}_{pre}/\nu^{273}_{pre}$ ratio of the neutron multiplicities values reported in Table I of paper [26], corresponding to the above-mentioned excitation energies for the two reactions (induced by the ^{26}Mg and ^{25}Mg beams respectively), the neutron emission probability $(\Gamma_n/\Gamma_{total})_1$ at first step of the ^{274}Hs CN, should be 72%, 77%, 85%, 88%, and 94%, respectively. This kind of result leads to the conclusion that for the ^{26}Mg+^{248}Cm reaction at $E^*_{CN}(^{274}$Hs) = 63 MeV of excitation energy, if the prescission neutron multiplicity ν^{274}_{pre} is a very low value as 1.7 or a very high value as 4, the probability to emit neutron at first step of the deexcitation cascade of the ^{274}Hs at E^*_{CN}=63 MeV is high as 72% - 94% against the fission that always should remain a shortly competitive process at decay of CN in comparison with the neutron emission, while on the contrary it is fission the dominant process at first step of the ^{274}Hs CN decay since the effective fission barrier B_f is 0.72 MeV in comparison with the binding neutron energy that is 6.51 MeV. Therefore, the method and procedures adopted in ref. [26] is clearly unreliable, and the result of formula (7) of ref. [26] is not sensitive to the reaction mechanism of the entrance channel.

We return again to discuss in the present paper about the Γ_n/Γ_{total} value of about 0.9 at E^*_{CN} =63 MeV found in the analysis of [26] for the ^{26}Mg+^{248}Cm reaction because the authors affirmed that such a relevant value of the probability of neutron emission Γ_n/Γ_{total} suggests the interpretation that the fission process of a high excited nucleus, to cause of the strong effect of the nuclear viscosity delays of reaching the saddle point, and only approaching about the last step of the deexcitation chain the intermediate excited nucleus can go into fission. Such a result of $(\Gamma_n/\Gamma_{total})_1$ and the related interpretation of delay of the fission process led to other inspections in experiments assuming that the result [26] and the competition of the neutron emission against fission is considered as a reliable result. In fact, in paper [28] the Γ_n value and its related interpretation were considered welcome with the words "Recently, it has been experimentally demonstrated that Γ_n is dominant over Γ_f in the first chance fission of ^{274}Hs produced in the ^{26}Mg+^{248}Cm reaction at E^* =63 MeV [57]."

But unfortunately the authors of [28] do not found results as they argued before in their investigations, and affirmed that "results should stimulate future measurements of both fission fragments and evaporation residues in reactions forming elements to understand fully the effects and properties of the slow quasifission that appears to be very important in ...." and finally they affirmed that " The development of parameter-free dynamical approaches to modelling heavy ion reactions is therefore desiderable for a full predictive capability on how best to produce heavy and superheavy elements and their isotopes."

IOP Conf. Series: Journal of Physics: Conf. Series **1014** (2018) 012004 doi:10.1088/1742-6596/1014/1/012004

5. Analysis of kinetic energy spectra of emitted light particles

As an example of sensitivity and possibility supplied by our code, we present in Fig.9 the calculated neutron energy spectra of some emitted neutrons at various steps of the deexcitation cascade of the $^{288}114$ CN* formed at $E_{\rm CN} = 35.89$ MeV of excitation energy in the ^{48}Ca+^{240}Pu reaction: thick full line represents the kinetic energy spectrum of neutron evaporated from excited $^{288}144$ CN, dashed line is for neutron emitted from excited $^{287}114^*$, dotted line is for neutron emitted from excited $^{286}144^*$, and thin full line is for neutron emitted from excited $^{285}114^*$.

At various steps of the deexcitation cascade of the formed CN, each intermediate excited nucleus surviving fission and reached after evaporation of one neutron, has a lower excitation energy E^* with respect to the $E^*_{\rm CN}$ value. Analogously to the various successive steps of the intermediate excited nuclei which survive fission along the deexcitation cascade, the yield of spectra of subsequent neutron emission is reduced of about two orders of magnitude and the correspondent kinetic energy range significantly decreases of about 5 MeV, respectively, after each neutron emission. In Fig.10 we present the calculated kinetic energy spectra of neutron, proton, and α-particle emitted when one of these particles is emitted at first step of the above-mentioned $^{288}114$ CN at excited energy $E_{\rm CN} = 40.06$ MeV. In this case the maximum yield of proton emission is reduced of about four orders of magnitude in comparison with the spectrum of the first neutron; for the emission of a charged particle from the internal part of CN exists a threshold for the evaporation of one charged particle. The maximum kinetic energy value of emitted proton is higher in comparison to the neutron emission to cause of the repulsive Coulomb force that appears on the charged particle after its emission from CN; in the case of the α-particle emission, the threshold energy is higher than the one for the proton emission and the kinetic energy range for the α-partice is much higher in comparison to the proton and neutron particle emissions.

In the present case of the ^{48}Ca+^{240}Pu reaction forming the $^{288}144$ CN, the intermediate excited nuclei reached along the deexcitation cascade are α-emitter and the maximum yield of α-emission is about 1 order of magnitude higher than the one of proton-emission.

Therefore, we can affirm that the experimental determinations of the kinetic energy spectra of α or/and proton emissions following the interaction of nuclei in the entrance channel can unambigously contribute to verify that the CN has really been formed with excitation energy $E^*_{\rm CN}$, and it is possible to observe the light charged particles evaporated from CN. Moreover, by also measuring the fusion-fission fragment cross sections vs. $E^*_{\rm CN}$ and the neutron kinetic energy distribution it is possible to determine the fusion cross section and to characterize the main relevant decay widths $\Gamma_{\rm n}/\Gamma_{\rm total}$, $\Gamma_{\rm fis}/\Gamma_{\rm total}$, $\Gamma_{\rm p}/\Gamma_{\rm total}$, and $\Gamma_\alpha/\Gamma_{\rm total}$ as functions of the excitation energy $E^*_{\rm CN}$

6. Conclusion

The reasons leading to uncertainties of the experimental and theoretical values of the fusion probability $P_{\rm CN}$ in heavy ion collision at lower energies are discussed. It should be stressed that there are two important reasons causing the uncertainties of the experimental values $P_{\rm CN}$. The first reason is related to the ambiguity in identification of the reaction products formed by the true capture and fusion events. In the analysis of experimental data with full momentum transfer, events with masses around the values of light initial nucleus and conjugate nucleus are not usually taken into consideration.

The second reason is the ambiguity in the separation of the fusion-fission events in the analysis of fission-like products containing quasifission or/and fast fission products. Therefore, the number of events seem to be larger than true fusion events due to consideration of the part of quasifission and fast fission events as fission events of compound nucleus which has not formed in the reaction. Certainly the extracted fusion probability $P_{\rm CN}$ will be larger than its correct

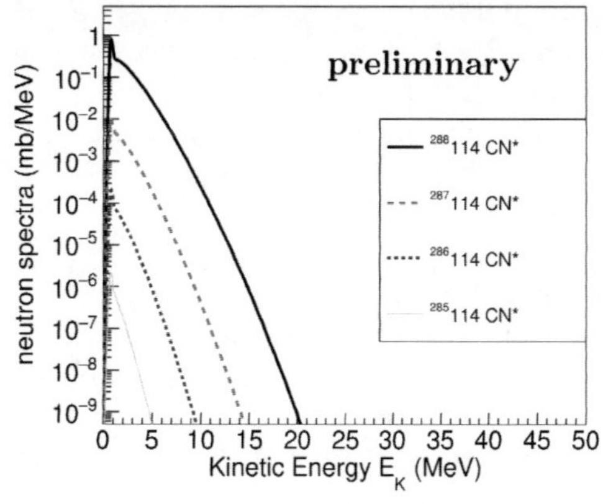

Figure 9. (Color on-line) Energy spectra of neutrons emitted at various steps of the deexcitation cascade starting from the ^{288}114 CN at E^*_{CN} = 35.89 MeV of excitation energy by the ^{48}Ca+^{240}Pu reaction: thick full line for the first neutron emitted from ^{288}114 CN*, dashed lines for the first neutron emitted from ^{287}114*, dotted line for the first neutron emitted from ^{286}114* and thin full line for the first neutron emitted from ^{285}114*.

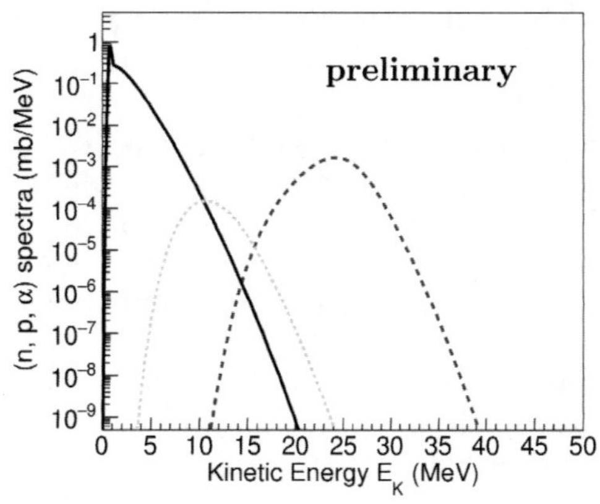

Figure 10. Energy spectra of n, p, α emitted from the ^{288}114 CN* at E^*_{CN} = 40.06 MeV of excitation energy by the ^{48}Ca+^{240}Pu reaction: full line for the first emitted neutron, dotted line for the first emitted proton, dashed line for the first emitted α-particle.

value.

Due to difference in the spin distribution probability of the heated and rotating nucleus the survival probability W_{sur} is strongly sensitive to the type of reactions in the entrance channel even if these reactions lead to the same CN formation with the same E^*_{CN} excitation energy.

The characteristics of our model and procedure used for calculation are based on the possibility to analyze the effect of the entrance channel on the nuclear reactions mechanism. Moreover, in the model all properties of reaction which are responsible for the evolution of reactants with formation of intermediate states and final products (mass asymmetry and type of deformations of interacting nuclei, potentials, barriers, excitation function, reaction mechanisms, competition of processes, cross sections, etc.) are considered as a function of the energy and angular momentum. We have shown that the consistent description of the fission cross section reached by consideration of the fade-out of the shell correction to the fission barrier with increasing temperature and angular momentum is very important for a reliable estimation of the final reaction products. This result is of crucial importance for the synthesis of the superheavy elements, since it extends the stabilizing effects of the shell structure to higher temperatures, but this stabilizing effect, however, will be partially removed by the decrease of the shell correction with the increasing angular momentum.

Moreover, in the present paper we propose to use the kinetic energy spectra of charged particles (α and/or proton) emission from the formed compound nucleus in order to obtain by an unambiguos way information on the characterization of formed excited nuclei. This methodology proves to be even more useful when, as the mass asymmetry of reacting nuclei in the entrance channel decreases, the reaction cross section is smaller than fraction of pb (picobarn) or even of fmb (femtobarn). In the case of reactants with $Z_1 \times Z_2$ >2500 the very low values of $P_{\rm CN}$ and $W_{\rm sur}$ functions lead to an estimation of evaporation residues lower than the fmb (femtobarn) to cause of high contributions of DNS decay into quasifission and to decay of CN into fission.

But, in any case, the possibility to form compound nuclei by collisions between nuclei with the $Z_1 \times Z_2$ value much higher than 5000 and by almost mass symmetric reactions it is a very unrealistic dream to obtain ERs because the reactants can not even reach the stage of capture. Therefore, without capture it is not possible to obtain fusion and then to reach the CN formation considered as a statistically equilibrated system from which it is possible to obtain ER nuclei by evaporation of light particles along the deexcitation cascade of CN.

Therefore the absence of a refined analysis with appropriate procedures as also the use of assumptions without estimation of effects on the final results can not contribute to a correct and useful understanding of the reaction mechanism that one wants to investigate.

References

[1] Nasirov A K *et al.* 2009 *Phys.Rev. C* **79** 024606
[2] Fazio G *et al.* 2005 *Mod. Phys. Lett. A* **20** 391
[3] Giardina G *et al.* 2010 *Int. J. Mod. Phys. E* **19** 882
[4] Nasirov A K *et al.* 2011 *Phys. Rev. C* **84** 044612
[5] Nasirov A K *et al.* 2013 *Eur. Phys. J. A* **49** 147
[6] Fazio G *et al.* 2008 *J. Phys. Soc. JPN.* **77** 124201
[7] Kim K et al. 2015 *Phys. Rev. C* **91** 064608
[8] Nasirov A K *et al.* 2010 *Journal of Physics: Conference Series* **205** 012018
[9] Mandaglio G *et al.* 2012 *Phys. Rev. C* **86** 064607
[10] Mandaglio G *et al.* 2015 *EPJ Web of Conferences* **96** 01016
[11] Giardina G *et al.* 2018 *Nucl. Phys. A* **970** 169
[12] Fazio G *et al.* 2004 *Eur. Phys. J. A* **22** 75
[13] Mandaglio G *et al.* 2012 *EPJ Web of Conferences* **38** 01001
[14] Fazio G *et al.* 2005 *Phys. Rev. C* **72** 064614
[15] Nasirov A K *et al.* 2005 *Nucl. Phys. A* **759** 342
[16] Sierk A J 1986 *Phys. Rev. C* **33** 2039
[17] Sobiczewski A and Pomorski K 2007 *Prog. Part. Nucl. Phys.* **58** 292
[18] Anastasi A *et al.* 2015 *Acta Physica Polonica B* **8** 583
[19] Ignatyuk A V *et al.* 1975 *Sov. J. Nucl. Phys.* **21** 255
[20] D'Arrigo A*et al.* 1994 *J. Phys. G* **20** 365
[21] Fazio G *et al.* 2003 *J. Phys. Soc. Jpn.* **72** 2509
[22] Ignatyuk A V *et al.* 1979 *Sov. J. Nucl. Phys.* **30** 626
[23] Hansen G and Jensen A S *Nucl. Phys. A* **406** 236
[24] Rastopchin E M *et al.* 1989 *Sov. J. Nucl. Phys.* **49** 15
[25] Satou K and Ikezoe H *et al.* 2002 *Phys. Rev. C* **65** 054602
[26] Yanez R *et al.* 2014 *Phys. Rev. Lett.* **112** 152702
[27] Dvorak J *et al.* 2008 *Phys. Rev. Lett.* **100** 132503
[28] Khuyagbaatar J Hinde D J *et al.* 2015 *Phys. Rev. C* **91** 054608

IOP Conf. Series: Journal of Physics: Conf. Series **1014** (2018) 012005 doi:10.1088/1742-6596/1014/1/012005

Multinucleon transfer reactions – a pathway to new heavy and superheavy nuclei?

Sophie Heinz[1]

[1] GSI Helmholtzzentrum, Planckstrasse 1, 64291 Darmstadt, Germany

s.heinz@gsi.de

Abstract. Recently, we reported the observation of several new neutron-deficient isotopes with proton numbers $Z \geq 92$ in collisions of ^{48}Ca + ^{248}Cm at the Coulomb barrier. The peculiarity is that these nuclei were produced in deep inelastic multinucleon transfer reactions, a method which is presently discussed as a possible new pathway to enter so far unknown regions in the upper part of the Chart of Nuclides. Of particular interest are multinucleon transfer reactions as a possible means to produce neutron-rich superheavy nuclei and nuclei along the magic neutron shell $N = 126$. Based on present-day physical and technical state-of-the art, we will discuss the question how big are our chances to enter these regions by applying multinucleon transfer reactions.

1. Introduction

New exotic nuclei are usually produced by fragmentation, by fission or in complete fusion reactions. The different regions of nuclei, which can be populated with these reactions are shown in figure 1. These established reactions, however, are exploited or not applicable, respectively, in the region of transuranium and superheavy nuclei, causing the necessity for novel methods to synthesize new heavy isotopes. In the last decade an intense discussion arose on the possibility to apply multinucleon transfer (MNT) reactions for the synthesis of new heavy and superheavy nuclei. One region of interest is neutron-rich transactinide and superheavy nuclei which cannot be produced in complete fusion reactions due to the lack of sufficiently neutron-rich projectile-target combinations. The other goal is to produce new isotopes in the region of astrophysical interest along the closed $N = 126$ neutron shell. Nuclei in this region are presently produced in fragmentation reactions at relativistic energies but model calculations predict larger cross-sections for new isotopes in this region if they are produced in MNT reactions. Multinucleon transfer reactions might indeed be an appropriate solution. They allow, in principle, to produce neutron-rich as well as neutron-deficient isotopes with proton numbers reaching far beyond uranium. Experimental and theoretical studies to produce heavy nuclei in MNT reactions at Coulomb barrier energies were already performed in the 1970s up to the 1990s [1-4]. The lowest cross-sections reached in some of those experiments were 20 nb. No new isotopes were observed. A revival of the topic was initiated about 10 years ago by new model calculations [5, 6] and measurements (see e.g. [7]) of reaction cross-sections and by the above named necessity to find methods different from the established fusion and fragmentation reactions. Meanwhile, the possible

IOP Conf. Series: Journal of Physics: Conf. Series **1014** (2018) 012005 doi:10.1088/1742-6596/1014/1/012005

application of MNT reactions for synthesis of new isotopes has become a topical subject in various laboratories and appropriate separation and detection techniques are in development.

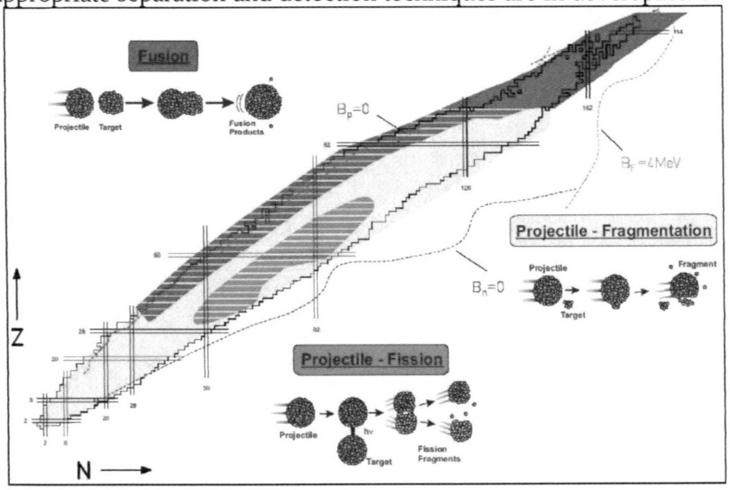

Figure 1. Chart of presently known nuclides and the reactions, which are typically applied to produce them [8].

Here, one has to keep in mind that the expected production cross-sections for new heavy and superheavy nuclei are small and reach far below the microbarn range. As a consequence, similarly efficient separation and detection techniques have to be applied as used for the identification of single atoms from fusion-evaporation reactions in superheavy element experiments. The techniques have also to be fast due to the short half-lives of exotic nuclei. Up to now no dedicated in-flight separators exist for heavy transfer products but the existing in-flight separators for heavy and superheavy evaporation-residues can also be applied for separation of transfer products. In particular, velocity filters are very sensitive to the reaction kinematics and allow for efficient separation between different isotopes. In recent years, we succeeded to produce for the first time new transuranium nuclei in MNT reactions [8] by using a highly sensitive experimental technique, namely, the velocity filter SHIP at GSI and its detection system.

2. Experimental technique

A sketch of the velocity filter SHIP and its detection system is shown in figure 2. Reaction products, which leave the target at angles of ±2 degree with respect to the beam direction are accepted by the entrance aperture of SHIP. All nuclei, which enter SHIP are separated according to their velocities. The accepted velocity window at a given setting is $\Delta v/v = 0.1$ (FWHM). All reaction products, which pass the velocity filter are implanted in a position sensitive 16 strip silicon detector (stop detector) where they are identified via α-decay tagging. The identification via α-decay chains is a very sensitive method, which enables in the utmost case the identification of single isotopes. In addition, a germanium clover detector is installed behind the stop detector to measure γ-rays emitted in prompt or delayed coincidence with particle registration in the stop detector. The fast in-flight separation allows, in principle, to detect short-lived nuclei with half-lives down to microseconds. This limit is given by the flight time of the reaction products through the 11 m long SHIP, which is about 1 μs. In our experiment, the shortest accessible half-lives were 20 μs given by the conversion time plus dead tine of the data acquisition system while the longest accessible half-lives are determined by the data recording time and the yield of a specific isotope.

Velocity filters are very sensitive to the reaction kinematics. In the same way as SHIP is typically used to separate heavy and superheavy fusion-evaporation residues, it can equally be applied to separate heavy target-like transfer products. Transfer products, which are scattered to the acceptance

IOP Conf. Series: Journal of Physics: Conf. Series **1014** (2018) 012005 doi:10.1088/1742-6596/1014/1/012005

angle of SHIP have velocities between 1.5 times and 2 times the compound nucleus velocity v_{CN}. Usually, a broad variety of transfer products is created in the reactions where each of the nuclei has a characteristic velocity depending on A and Z and the energy dissipation during the reaction. In order to cover the range of different velocities expected in an experiment, we varied stepwise the velocity setting of SHIP and recorded at each setting the α- and γ-decays of the implanted reaction products and their daughter nuclei.

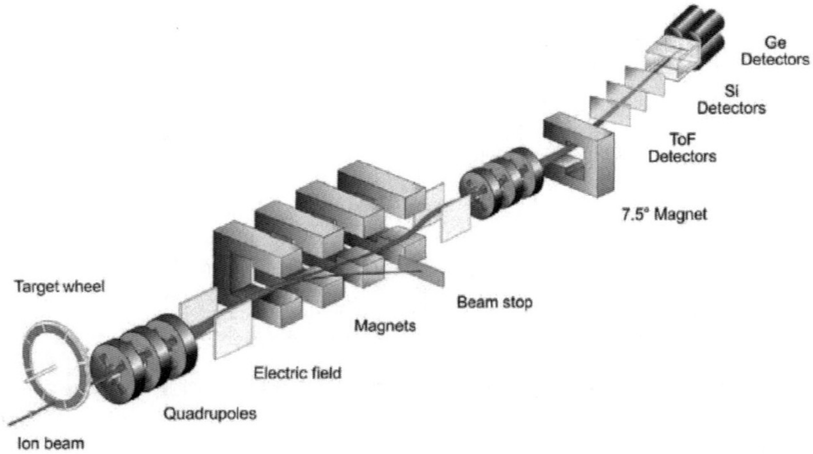

Figure 2. Sketch of the experimental setup: the velocity filter SHIP and the detection system.

In reactions at Coulomb barrier energies, the created MNT products have wide angular distributions, but due to the narrow acceptance angle of SHIP, only nuclei which are emitted in forward direction are accepted which results in typical efficiencies for MNT products between 0.1% and 1%. The selection of nuclei in forward direction means at the same time selection of nuclei from central collisions where projectile and target nucleus reach maximum possible overlap and sticking time at the given beam energy. This is the most favourable configuration for the exchange of large numbers of nucleons leading to MNT products far from the original projectile and target nuclei.

3. Experimental results

In reactions of ^{48}Ca + ^{248}Cm at 5.3 MeV/u we observed about 100 different nuclides, which were populated in MNT reactions (see grey squares in figure 3). The isotope identification was performed via the α-decay properties of the nuclei. Usually, a contiguous region of isotopes around the target and projectile nucleus is populated in transfer reactions. However, the nuclei, which are represented by blank squares in figure 3 could not be detected in our experiment due to too short or too long half-lives and / or unfavourable decay channels like fission or β-decay. Apart from a large number of already known nuclides, our data revealed also some new isotopes (marked by dots), which are located in the region Z ≥ 92 on the neutron-deficient side of the Chart of Nuclides. All of them lead to decay chains of already known nuclei which enabled their assignment. For a detailed description of the experiment and results we refer to [10].

The cross-sections of the observed isotopes for elements Z = 84 – 95 are shown in figure 4 as a function of their mass number A, in the following called isotopic distributions. The cross-sections are related to the cumulative number of α-decays detected at (0 ± 2) degree. This means, they comprise for each isotope nuclei which were directly produced in the reaction as well as nuclei populated by precursor decays. The lowest measured cross-sections were on the level of 10 pb/sr, which corresponds to an average event rate of one nucleus per day in the stop detector at the given beam intensity of $2 \cdot 10^{12}$ projectiles / s and target thickness of 460 μg/cm^2.

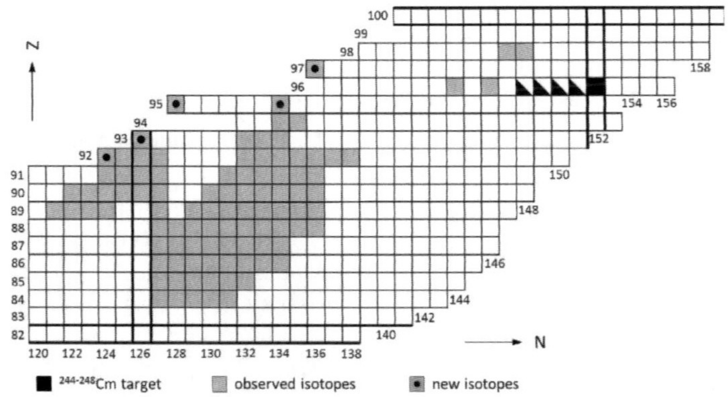

Figure 3. Survey of all multinucleon transfer products which we identified in reactions of ^{48}Ca + ^{248}Cm at the beam energy 5.3 MeV/u (grey squares). Beside ^{248}Cm with a contribution of 96.85% (black square) the target contained also traces of the isotopes $^{244-247}$Cm (black triangles). The five new isotopes, which we published in [9] are marked by dots.

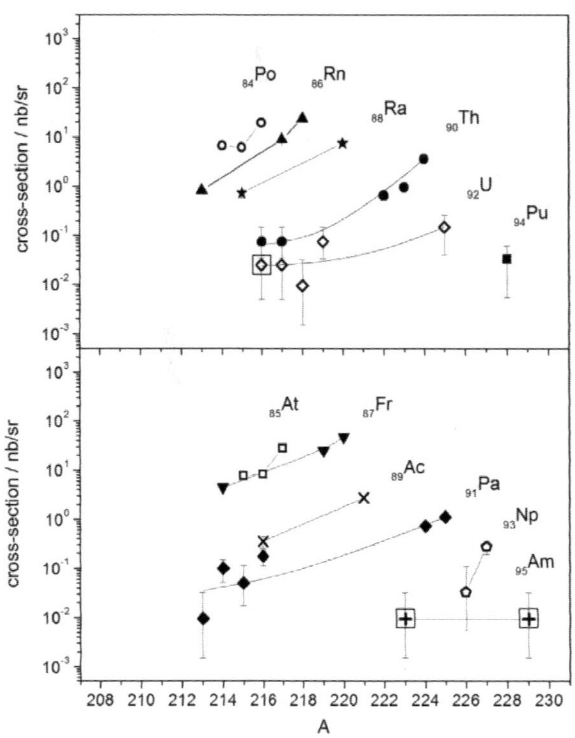

Figure 4. Isotopic distributions of multi-nucleon transfer products with Z = 84 – 95 measured in collisions of ^{48}Ca + ^{248}Cm at 5.3 MeV/u. The cross-sections are related to the cumulative number of α-decays which comprise decays of directly populated nuclei and of nuclei populated by precursor decays. Upper figure: distributions for even-Z nuclei; lower figure: distributions for odd-Z nuclei. The new isotopes are framed. The lines are drawn to guide the eye.

4. Discussion

The results of our experiment reveal that the application of a velocity filter in combination with α-decay tagging for the identification of the reaction products allows to reach in few days of beamtime cross-sections on the level of 10 pb/sr. According to model predictions, this is, in principle, already sufficient to reach new isotopes in the envisaged regions named above. In the following, we are going to shed more light on the possible application of MNT reactions to populate these regions.

4.1 Neutron-deficient transuranium isotopes

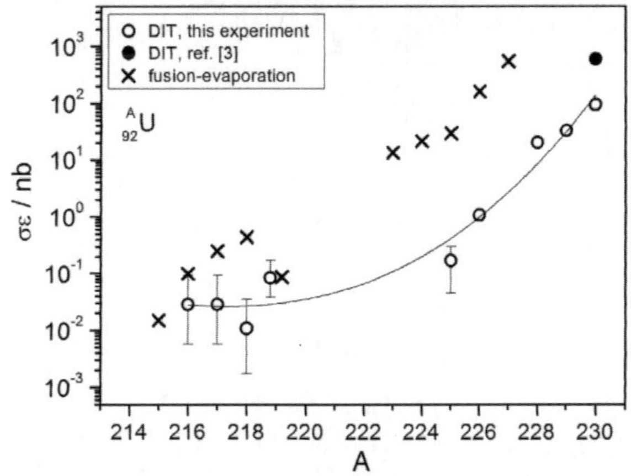

Figure 5. Product of cross-section and experimental efficiency, σε, for uranium isotopes measured in fusion-evaporation reactions (crosses) and in multinucleon transfer reactions of ^{48}Ca + ^{248}Cm in our experiment (open circles) and in the experiment in ref. [3] (full circle). For the transfer products we show cumulative cross-sections. For ^{219}U we put for better visibility a small offset on the A value to avoid the overlap of the data points from transfer and fusion.

The established method to produce neutron-deficient transuranium isotopes is fusion-evaporation reactions. Therefore, it is interesting to compare them with MNT reactions concerning the capacities of the two methods for isotope production. As a starting point we show in figure 5 measured fusion-evaporation cross-sections for various uranium isotopes [11 - 17] and MNT cross-sections measured in our experiment and in [3]. Figure 5 can be regarded as a representative example, because the behaviour of the cross-sections is similar for other elements in the uranium region. The fusion cross-sections for the same residual nucleus can vary by orders of magnitude depending on the applied projectile-target combination and beam energy. If cross-sections from more than one experiment were available for the same isotope, we put the largest known value in figure 5. Concerning MNT reactions, it is very likely that the given cross-sections have not the maximum achievable values, because systematic studies to find the optimum projectile-target combinations and measurements of excitation functions for the given MNT products were not performed yet.

Instead of the pure reaction cross-sections we show in figure 5 the product of cross-section and experimental efficiency, σε, because this is the experimentally relevant parameter which reflects the event count rate. In our case, ε is mainly determined by the angular acceptance of the applied separators for fusion and transfer products, respectively. The following trend is noticed in figure 5: for the more neutron-rich isotopes well above A = 220, σε is significantly larger in fusion-evaporation reactions. This is on one hand caused by the significantly larger angular acceptance of the separators for fusion-evaporation residues. Another reason is that these isotopes can be produced with very asymmetric projectile-target combinations like ^{22}Ne + ^{208}Pb → ^{230}U* where the low entrance channel Coulomb barriers and favourable Q-values lead to relatively large fusion residue cross-sections. In this region, fusion-evaporation reactions appear superior to MNT reactions. For very neutron-deficient isotopes the situation changes and σε for fusion and MNT products tend to approximate each other despite the small efficiencies for MNT products. To reach these nuclei in fusion reactions, more symmetric projectile-target combinations have to be applied like e.g. ^{40}Ar + ^{182}W → ^{222}U*. The larger entrance channel Coulomb barriers and less favourable Q-values lead to relatively small fusion-evaporation residue cross-sections. In this region, a clear advantage of MNT reactions comes into play, namely, their broad excitation functions. Resulting from this, a vast region of isotopes can be populated and detected simultaneously at a given beam energy while fusion-evaporation reactions are only selective on few isotopes at a given experimental setting. Therefore, MNT reactions appear indeed as an attractive way for the production of new neutron-deficient transuranium isotopes.

4.2 Neutron-rich superheavy nuclei

There is presently no experimental method to produce neutron-rich nuclei well beyond uranium. Fragmentation reactions are not applicable because the heaviest available projectile or target nucleus, respectively, is uranium and with fusion reactions only neutron-deficient heavy nuclei can be produced due to the bending of the stability line toward the neutron axis with increasing Z. Therefore, MNT reactions are presently the only potential method to access these nuclei. The sensitivity reached in our experiment on $^{48}Ca + ^{248}Cm$ allows already, in principle, to enter the region of neutron-rich nuclei with Z > 100 where cross-sections in the nanobarn region are expected from model calculations for new neutron-rich isotopes around Z = 102. The main bottleneck, which presently prevents the observation of MNT products in this region is the lack of appropriate identification techniques. Alpha decay tagging is not applicable because the expected dominant decay mode of neutron-rich superheavy nuclei is spontaneous fission. Moreover, also the half-lives can become too long for decay tagging in this region. Also the determination of A and Z by measurement of energy, energy loss and time-of-flight is not possible because at energies around the Coulomb barrier plasma effects and the resulting pulse height deficit do not allow to determine A and Z of very heavy nuclei with an accuracy of one unit.

A satisfactory solution for this problematic is not yet existing. There are different approaches for new identification techniques which are, however, not satisfactory, mainly from terms of their present efficiency. One approach is to determine the nuclear charge number by selective laser ionization while the mass number of the singly ionized nuclei can be obtained with a magnetic dipole field. Another approach is the application of precision mass measurements with a Penning trap or multiple reflection time-of-flight mass spectrometer where a resolving power of 10^5 – 10^6 is already sufficient for an isobaric separation of most of the heavy nuclei. A third approach, especially for A identification, is the application of E-TOF telescopes where calorimeters are used for energy measurement instead of silicon detectors. Calorimetric low temperature detectors were already studied for slow heavy ions and allow for a resolution of A close to one unit even for nuclei in the uranium region [18]. However, all these methods lead presently to tiny detection efficiencies, which are far from allowing the detection of single nuclei.

4.3 Nuclei along the N=126 neutron shell

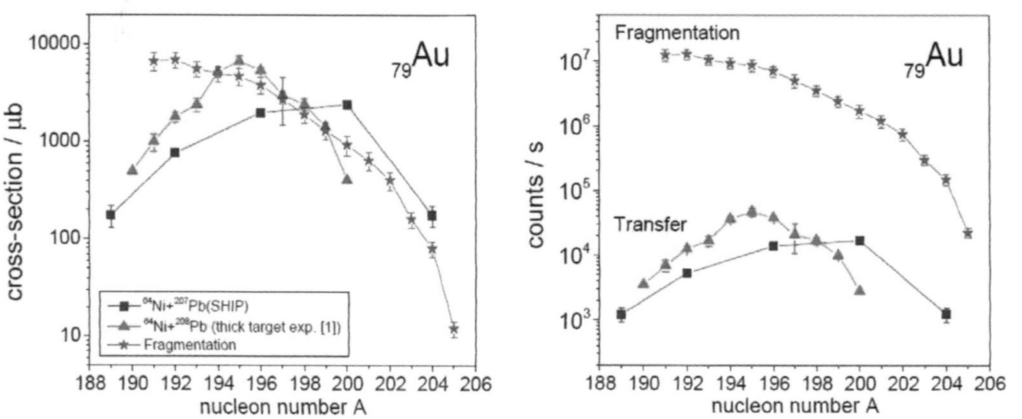

Figure 6. Production cross-sections and expected yields (at the target) for Au isotopes produced in multinucleon transfer reactions of $^{64}Ni + ^{207,208}Pb$ and in fragmentation reactions.

In collisions of $^{64}Ni + ^{207}Pb$ at 5.0 MeV/u we studied the population of isotopes below Pb in MNT reactions [19]. The nuclides below Pb are β^- emitters (on the neutron-rich side of the stability line) or undergo electron capture decay (neutron-deficient side). Therefore, we used delayed γ-rays emitted by the reaction products, which are implanted in the stop detector, for isotope identification. This method is less sensitive than α-decay tagging and lead to lower cross-section limits of 10 μb//sr. We identified several isotopes of elements Z = (76 – 81). As a representative example, we show in figure 6 the results for Au (Z = 79) isotopes (full squares). Similar results were obtained 10 years earlier by another group in a thick-target experiment performed with collisions of ^{64}Ni + ^{208}Pb [20]. The results of this experiment are shown by triangles in figure 6.

The left part of figure 6 shows the production cross-section of Au isotopes as a function of their nucleon number A. Our cross-sections and the ones measured in [20] differ in most cases within one order of magnitude. The data from MNT reactions of both experiments indicate a rather steep downward slope toward the neutron-rich side after the maximum is reached around nuclei located close to the stability line. In both experiments no new isotopes were observed and as well the N = 126 shell was not reached for nuclei with Z < 82. Since the presently applied technique to produce neutron-rich isotopes below Pb is projectile fragmentation, it is interesting to compare the isotopic distributions and cross-sections measured in MNT reactions with those from fragmentation reactions [21, 22]. The fragmentation data are represented by asterisks. Also the fragmentation cross-sections show a steep decrease towards neutron-rich nuclei but the experiments are more sensitive and allowed to reach cross-sections below 10 nb. The fragmentation and MNT cross-sections are mostly within the sane order of magnitude for the isotopes discussed here. We could notice the trend that the MNT cross-sections are going to exceed the ones from fragmentation if one goes further below Pb, which is well in agreement with theoretical predictions.

Table 1. Typical parameters for beam intensities N_{beam}, target thicknesses d_{target} and angular efficiencies ε in multinucleon transfer and in fragmentation reactions.

	N_{beam}	d_{target}	ε
Transfer	$5 \cdot 10^{12} / s$	$500\ \mu g / cm^2$	< 5% (SHIP)
Fragmentation	$5 \cdot 10^{9} / s$	$5\ g / cm^2$	< 50% (FRS)

Next, we are going to compare the production *yields* for the observed isotopes from MNT and fragmentation because this is finally the decisive criterion for the successful application of a reaction type for isotope production. To deduce yields from the crtoss-sections, one has to take into account the different conditions concerning beam intensities, target thickness and detection efficiencies for MNT and fragmentation products. Typical parameters are shown in table 1. The yields (at the target) for Au isotopes are shown on the right side of figure 6. One can see that the estimated yields for the same isotopes are one or more orders of magnitude larger in fragmentation reactions. If we regard in addition the expected yields at the detector, the difference between the two production methods is still increasing in favor of fragmentation reactions caused by the very different angular distributions of MNT and fragmentation products. In projectile fragmentation reactions the nuclei are emitted in a very narrow forward cone due to the relativistic energies and inverse kinematics. Resulting from this, very large angular efficiencies of nearly 100% can be reached despite of the small acceptance angles of the separators. MNT products, in contrast, are emitted in a broad angular window of typically 50 degree or more which makes an effective application of in-flight separators difficult. Finally, another point has to be mentioned. In fragmentation reactions, the identification of the isotopes is performed by measuring energy, energy loss, time-of-flight and magnetic rigidity of the nuclei. Due to the relativistic energies, A and Z of the nuclei can be resolved with one unit. In MNT reactions at

Coulomb barrier energies, plasma effects and the resulting pulse height deficit in the silicon detector limit the A and Z resolution, which is usually not better than three units in the region around Pb.

These results and considerations show that fragmentation reactions appear much more favorable for the production of neutron-rich nuclei along the N = 126 shell. And based on present day knowledge, this will be also valid for future experiments, in particular after the upgrades of beam intensities which are ongoing in present radioactive ion beam facilities.

5. Summary and Outlook

As outlined in the previous sections, MNT reactions in heavy collision systems are presently discussed as a possible new pathway to produce new isotopes in the following regions on the upper part of the Chart of Nuclides: (i) neutron-rich transuranium and superheavy nuclei, (ii) neutron-deficient transuraniun nuclei and (iii) neutron-rich nuclei with Z < 82 along the N=126 neutron shell. What can we conclude for the success of this enterprise? Based on the presently available, still rather scarce experimental data one can say that there is no overall valid answer concerning the three regions of interest. Rather, it seems that we have to treat each region individually and arrive at quite different conclusions. The answer is rather optimistic concerning the production of very neutron-deficient transuranium isotopes. Here it even seems that MNT reactions might be more profitable compared to the presently applied fusion reactions. The expected yields from MNT are comparable to the ones from fusion reactions, but a big advantage of MNT reactions is that a wide region of isotopes can be populated in the same experiment while fusion-evaporation reactions are selective on only few isotopes at given beam energy and experimental setting. This is due to the broader excitation functions of MNT products.

For the production of neutron-rich transuranium nuclei, the cross-sections obtained in MNT reactions seem at least sufficient to produce isotopes of elements up to $Z \approx 105$. However, here appears the bottleneck of missing identification techniques for single heavy nuclei. The only existing powerful method of alpha decay tagging cannot be applied because nuclei in this region are very long-living and mostly do also not decay via alpha emission. Nevertheless, MNT reactions seem presently the only possibility to enter the region of neutron-rich transuranium and superheavy nuclei. Therefore, the first and decisive step is the development of new and powerful detection methods, which allow the identification of these nuclei independent of their decay properties.

For the third region of nuclei along N=126, Z < 82 the answer seems quite sure that projectile fragmentation reactions will also in the future remain the most powerful method for their production. Even if the production cross-sections in fragmentation and MNT turned out to be similar, the yields in fragmentation reactions are several orders of magnitude higher because of the more favourable conditions concerning target thickness and experimental efficiencies. Further, in fragmentation reactions the problem of isotope identification does not appear like in MNT reactions. Due to their relativistic energies, A and Z of the fragmentation products can be determined by measuring energy, energy loss, time-of-flight and magnetic rigidity.

References

[1] Freiesleben H *et al.*, 1979 Z. *Phys. A* **292** 171
[2] Schädel M *et al.* 1982 *Phys. Rev. Lett.* **48** 852
[3] Gäggeler H *et al.* 1986 *Phys. Rev. C* **33** 1983
[4] Türler A *et al.* 1992 *Phys. Rev. C* **46** 1384
[5] Adamian GG, Antonenko NV and Zubov AS 2005 *Phys. Rev. C* **71** 036603
[6] Penionzhkevich Yu E, Adamian GG and Antonenko NV 2005 *Phys. Lett. B* **621** 119
[7] Corradi L *et al.*, 2002 *Phys. Rev. C* **66** 024606

IOP Conf. Series: Journal of Physics: Conf. Series **1014** (2018) 012005 doi:10.1088/1742-6596/1014/1/012005

[8] Knöbel R priv. comm.
[9] Devaraja HM *et al.* 2015 *Phys. Lett. B* **748** 199
[10] Heinz S *et al.* 2016 *Eur. Phys. J. A* **52** 278
[11] Yang HB *et al.* 2015 *Eur. Phys. J. A* **51** 88
[12] Malyshev ON *et al.* 2000 *Eur. Phys. J. A* **8** 295
[13] Leppänen AP *et al.* 2007 *Phys. Rev. C* **75** 054307
[14] Andreyev AN *et al.* 1991 *Z. Phys. A* **338** 363
[15] Heßberger FP *et al.* 1989 *Z. Phys. A* **333** 111
[16] Yeremin AV *et al.* 1994 *Nucl. Intrum. Methods B* **350** 608
[17] Kalaninova Z *et al.* 2015 *Phys. Rev. C* **92** 014321
[18] Kraft-Bermuth S *et al.* 2009 *Rev. Sci. Intrum.* **80** 103304
[19] Beliuskina O *et al.* 2014 *Eur. Phys. J. A* **50** 161
[20] Krolas W *et al.* 2003 *Nucl. Phys. A* **724** 289
[21] Casarejos E *et al.* 2006 *Phys. Rev. C* **74** 044612
[22] Teresa Kurtukian-Nieto, PhD thesis work, University of Santiago de Compostela, Spain, 2007.

IOP Conf. Series: Journal of Physics: Conf. Series **1014** (2018) 012006 doi:10.1088/1742-6596/1014/1/012006

Nuclear Reactions Studies in Laser-Plasmas at the forthcoming ELI-NP facilities

G. Lanzalone[1,2,*], A. Muoio[1,4], C. Altana[1,3], M. Frassetto[5], L. Malferrari[5], D. Mascali[1], F. Odorici[5], S. Tudisco[1], L. A. Gizzi[6,7], L. Labate[6,7], S. M. R. Puglia[1], A. Trifirò[4,8]

[1] Istituto Nazionale di Fisica Nucleare - Laboratori Nazionali del Sud – Via S. Sofia 62, 95123 Catania, Italy
[2] Università degli Studi di Enna "Kore" – Via delle Olimpiadi, 94100 Enna, Italy
[3] Dipartimento di Fisica e Astronomia, Università degli Studi di Catania – Via S. Sofia 64, 95123 Catania, Italy
[4] Dipartimento di Scienze MIFT, Università degli Studi di Messina – Viale F.S. D'Alcontres 31, 98166 Messina, Italy
[5] Istituto Nazionale di Fisica Nucleare – Sezione di Bologna – Viale B. Pichat 6/2, 40127 Bologna, Italy
[6] Istituto Nazionale di Ottica CNR - Via G. Moruzzi, 1 – 56124 Pisa, Italy
[7] Istituto Nazionale di Fisica Nucleare, Sezione di Pisa, Largo Bruno Pontecorvo 3 - 56127 Pisa, Italy
[8] Istituto Nazionale di Fisica Nucleare, Sezione di Catania - Via S. Sofia 64, 95123 Catania, Italy
*Corresponding author e-mail : Gaetano.Lanzalone@lns.infn.it

Abstract. This work aim to prepare a program of studies on nuclear physics and astrophysics, which will be conducted at the new ELI-NP Laser facility, which actually is under construction in Bucharest, Romania. For the arguments treated, such activity has required also a multidisciplinary approach and knowledge in the fields of nuclear physics, astrophysics, laser and plasma physics join with also some competences on solid state physics related to the radiation detection. A part of this work has concerned to the experimental test, which have been performed in several laboratories and in order to study and increase the level of knowledge on the different parts of the project. In particular have been performed studies on the laser matter interaction at the ILIL laboratory of Pisa Italy and at the LENS laboratory in Catania, where (by using different experimental set-ups) has been investigated some key points concerning the production of the plasma stream. Test has been performed on several target configurations in terms of: composition, structure and size. All the work has been devoted to optimize the conditions of target in order to have the best performance on the production yields and on energies distribution of the inner plasma ions. A parallel activity has been performed in order to study the two main detectors, which will constitute the full detections system, which will be installed at the ELI-NP facility.

1. Introduction

Given its nature, the plasma state is characterized by a complexity that far exceeds the one exposed by the solid, liquid, and gaseous states. Correspondingly, the physical properties of nuclear matter (structure, reaction mechanisms, lifetimes, etc.) could change inside the plasma. Thus, the study of

these properties represents one of the most far ranging, difficult and challenging research areas today. Implications can cover also others fields, from quantum physics to cosmology, astrophysics, etc.
One of the crucial topics related to nuclear reactions in the ultra-low energy regime is the electron screening, which prevents a direct measurement of the bare nucleus cross section at the energies of astrophysical interest. Since in the laboratory interacting particles are in the form of neutral atoms, molecules or ions, in direct experiments at very low beam energy, electron clouds partially screen the nuclear charges thus reducing the Coulomb suppression. This results in an enhancement of the measured cross section compared with the bare nucleus one. The electron screening effect is significantly influenced by the target conditions and composition. In this context, it is of particular importance the measurement of cross-sections at extremely low energetic domains including plasmas effect, i.e. in an environment that under some circumstances and assumptions can be considered as "stellar-like". A further key point is connected with the fact that in such environment nuclear reactions can be triggered also by the excited states of the interacting nuclei. Thus, determining the appropriate experimental conditions that set the role of the excited states in the stellar environment can strongly contribute to the development of nuclear astrophysics [1]. The study of direct measurements of reaction rates in plasma offers this chance.
The future availability of high-intensity laser facilities [2,3] capable of delivering tens of peta-watts of power into small volumes of matter at high repetition rates will give the unique opportunity to investigate nuclear reactions and fundamental interactions under extreme plasma conditions, including also the influence of huge magnetic and electric fields, shock waves, intense fluxes of γ and X rays originating during plasma formation and expansion stages.
A laser is a unique tool to produce plasma and very high fluxes of photons and particles beams in very short duration pulses. Both aspects are of great interest for fundamental nuclear physics studies. In a plasma, the electron–ion interactions may modify atomic and nuclear level properties. This is of prime importance for the population of isomeric states and for the issue of energy storage in nuclei. Nuclear properties in the presence of very high electromagnetic fields, nuclear reaction rates or properties in hot and dense plasmas are new domains of investigation. Furthermore, with a laser it is possible to produce electric and magnetic fields strong enough to change the binding energies of electronic states. If nuclear states happen to decay via internal conversion (IC) through these perturbed states, a modification of their lifetimes will be seen. The excitation of nuclear levels by means of energy transfer from the atomic part to the nuclear part of an atom is the subject of a large number of investigations. Their goal is to find an efficient mechanism to populate nuclear isomers in view of further applications to energy storage and development of lasers based on nuclear transitions. In addition, other new topics can be conveniently explored such as three-body fusion reactions as those predicted by Hoyle [4,5].
Several Laser facilities are under construction around the world to push the physics beyond the actual level of knowledge. Among of these, the Extreme Light Infrastructure for Nuclear Physics at Magurele (Bucharest) in Romania, will be the only one devoted to Nuclear Physics studies.
ELI-NP [2] will be made up of a very high intensity laser system, consisting of two 10 PW laser arms able to reach intensities of 10^{23} W/cm2 and electrical fields of 10^{15} V/m, and very short wavelength γ beams with very high brilliance (10^{13} γ/s) and energy up to 19.5 MeV. This combination allows for three types of experiments: stand-alone high power laser experiments, stand-alone γ beam experiments and combined experiments of both facilities. Here the low repetition rate (1/min) of the high power laser requires the same low repetition rate for the γ beam in combined experiments. While the standalone γ beam will be used with typically 120 kHz, the low repetition mode requires few very intense γ pulses. With the high power laser we do not plan to interact with nuclear dynamics directly, but we use the laser for ion acceleration or to produce relativistic electron mirrors followed by a coherent reflection of a second laser beam in order to generate very brilliant X-ray or γ beams. We plan to use these beams later to produce exotic nuclei or to perform new γ spectroscopy experiments in the energy or time domain. The production of heavy elements in the Universe, a central question of astrophysics, will be studied within ELI-NP in several experiments.
In this article some of the activities, related to the project of study of nuclear astrophysics at ELI-NP will be discussed. A brief introduction is given to present the main open problems on nuclear

astrophysics and the opportunity offered by the laser matter interaction scheme. Afterwards, it is shown a short presentation of ELI-NP project and the studies performed to prepare the future activities and test on plasma and nuclear detectors. Finally, same results of the tests performed on the new targets are presented and discussed.

2. Open problems on nuclear astrophysics

Accurate measurements of nuclear reaction rates of proton and alpha burning processes are essential for the correct understanding of many astrophysical processes, such as stellar evolutions, supernova explosions and Big Bang nucleo-synthesis, etc. To this aim direct and indirect measurements of the relevant cross sections have been performed over the years. Since in the laboratory interacting particles are in the form of neutral atoms, molecules or ions, in direct experiments at very low beam energy, electron clouds partially screen the nuclear charges thus reducing the Coulomb suppression [6]. This results in an enhancement of the measured cross section compared with the bare nucleus one [7]. The electron screening effect is significantly affected by the target conditions and composition [8]; it is of particular importance the measurement of cross-sections at extremely low energetic domains including plasmas effect, i.e. in an environment that under some circumstances and assumptions can be considered as "stellar-like" (for example, for the study of the role played by free/bounded electrons on the Coulombian Screening can be done in dense and warm plasmas). Electron screening prevents a direct measurement of the bare nucleus cross section at the energies of astrophysical interest. In the last decade, the bare cross section has been successfully measured in certain cases by using several indirect methods [9]. Habitually, astrophysical appropriate reactions are performed in laboratories with both target and projectile in their ground state. However, at temperatures higher than about 10^{8}K, an important role can be also played by the excited states, as already deeply discussed in the innovative theoretical work of Bahcall and Fowler [10]. In that case, the authors studied the influence of low lying excited ^{19}F states on the final ^{19}F(p,alpha) reaction, predicting an increase of a factor of about 3 in reaction rate at temperatures of about 1-5 GK.
Thus determining the appropriate experimental conditions that allow the role of the excited states in the stellar environment could strongly contribute to the development of nuclear astrophysics. The study of direct measurements of reaction rates in plasma offers this chance. In addition others new topics can be conveniently explored such as three body fusion reactions as those predicted by Hoyle [4], lifetime changes of unstable elements [11] or nuclear and atomic levels [12] in different plasma environments; other fundamental physics aspects like non-extensive statistical thermodynamics [13] can be investigated in order to validate/confute the general assumption of local thermal equilibrium that is traditionally done for plasmas. The future availability of high-intensity laser facilities capable of delivering tens of peta-watts of power (e.g. ELI-NP) into small volumes of matter at high repetition rates will give the unique opportunity to investigate nuclear reactions and fundamental interactions under extreme conditions of density and temperature that can be reached in laser generated plasmas [14], including the influence of huge magnetic and electric field, shock waves, intense fluxes of X and γ-ray originated during plasma formation and expansion stages.
To investigate these research topics, we are proposing the construction of a general-purpose experimental set-up, where it will be possible to study the electronic screening problem in a wide variety of cases and configurations with different purposes [4,5,15]. Through the laser-target interaction, we aim at producing plasmas containing mixtures of ^{13}C + ^{4}He and ^{7}Li + d appropriate to investigate inner-plasma thermo-nuclear reactions. The ^{13}C+^{4}He reaction is of key interest for the investigation of the helium burning process in advanced stellar phases [16]. The ^{7}Li(d,n)^{4}He-^{4}He reaction was recently addressed by Coc et al. [17] as one of the most important reactions affecting the CNO abundances produced during the primordial nucleosynthesis (BBN). Very few experimental data exist, and authors consequently assume a constant S-factor ranging between two extreme hypotheses values.
Also, we propose to investigate the ^{11}B(^{3}He, d)^{12}C* reaction in a plasma. Nucleonic matter displays a quantum-liquid structure, but in some cases finite nuclei behave like molecules composed of clusters of protons and neutrons. To perform the proposed experiments, providing relevant data concerning the aforementioned reactions and others, we aim to take advantage from the excellent and unique

performance of the ELI-NP [2] facility and realize an experimental setup where two laser beams generate two colliding plasmas. The reaction products (neutrons and charged particles) will be detected through a new generation of plastic scintillators wall and through a new silicon carbides wall. The use of colliding plasma plumes appropriate for nuclear physics studies was suggested few years ago at LNS [14] and recently adopted also by other research teams [18]. One of the possible schemes may be the following: a first laser pulse imping on a ^{13}C, ^{7}Li or ^{11}B solid thin target (few micro-meters) producing, through the well-known TNSA [19] (Target Normal Sheath Acceleration) acceleration scheme, boron, carbon or lithium plasma. The rapidly streaming plasma impacts on a secondary plasma, prepared through the interaction of a second laser pulse on a gas jet target (made by ^{4}He, D_2 or ^{3}He). TNSA [19] was intensively studied in the last years; experiments [20] and models [21] show that this acceleration scheme works very well in the intensity domain between 10^{18}-10^{20} W/cm^2. The produced ions expand along a cone, whose axis is normal to the target surface, with a relatively low emittance [22].

3. Experimental Set-Up

The project will take advantage from the excellent and unique performance of the ELI-NP facility to create two colliding plasmas using two separate laser beams. The use of colliding plasma plumes suitable for nuclear physics studies [23] was proposed few years ago [14] and recently adopted to achieve such goal [18]. The idea is the following: a first laser pulse impinging on a 13C, 7Li or 11B solid thin target (few micro-meters) produces, through the TNSA (Target Normal Sheath Acceleration) [20] acceleration scheme, boron, carbon or lithium plasma. In view of this, an extensive experimental investigation programme is in progress, aiming at the optimization of targets and interaction configuration [24,25]. The rapidly streaming plasma impacts on a secondary plasma, prepared through the interaction of a second laser pulse on a gas jet target (made by 4He, D2 or 3He). The proposed activity requires also the construction of a highly segmented detection system for neutrons and charged particles. The segmentation is required for the reconstruction of the reactions kinematic. The ideal neutron detection module for these studies must exhibit: high efficiency, good discrimination of gammas from neutrons, good timing performance for ToF (Time of Flight) neutron velocity reconstruction. In addition, it must be able to work in hard environmental conditions, like the ones established in the laser-matter interaction area. All these aspects may be met by configuration based on 50x50x50 mm^3 PPO-Plastic scintillator plus a SiPM [26] read-out and a totally digital acquisition of the multi-hit signals. Concerning the charged particle detector, an R&D activity was funded by INFN on Silicon Carbide [27]. In this framework we aim to realize a wall of detectors. The SiC detectors have been proven recently to have excellent properties [28,29]: high energy and time resolution, resistance to radiation, low sensitivity to visible light, etc. The use of segmented SiC detectors would be very helpful for the study of nuclear reactions where only the position and energy measurement of light charged particles can give access to the desired information. In conclusion, we present in this contribution the research project that our collaboration will conduct in the next years @ELI-NP. More details about these activities are reported in reference [2].

4. Experimental Investigation on acceleration mechanisms and target.

A experimental campaign aiming at investigating the ion acceleration mechanisms through laser-matter interaction in femtosecond domain has been carried out at the Intense Laser Irradiation Laboratory (ILIL-Pisa, Italy) [30] facility with a laser intensity of up to 2 10^{19}W/cm^2. In recent years, laser ion acceleration has gained much interest focusing on fast ions emitted from a solid target by intense laser irradiation [31,32]. Protons and heavier ions can be accelerated up to tens MeV per nucleon via various mechanisms such as target normal sheath acceleration (TNSA) [19,20,33,34], radiation pressure acceleration (RPA) [35-39], and break-out afterburner acceleration (BOA) [40-42]. Our work concerns the TNSA regime, in which an intense laser pulse is focused ($> 10^{18}$ W/cm^2) on different micrometers thick foil target.

In this context, we carried out a systematic experimental investigation to identify the role of target properties in TNSA, with special attention to target thickness and dielectric properties. We used a full range of ion, optical, and X-ray diagnostics to investigate laser-plasma interaction and ion

IOP Conf. Series: Journal of Physics: Conf. Series **1014** (2018) 012006 doi:10.1088/1742-6596/1014/1/012006

acceleration. We focus on the results obtained using a Thomson Parabola Spectrometer (TPS). In the TPS, often used in such experiments, ions with different charge-to-mass ratios are separated into distinct parabolas. This allows to extract information for each ion species when several ions are generated simultaneously in a given solid angle. In the paper [43] we discuss the energy spectra of light-ions depending on structural characteristics of the target. Surface and volume contributions to the ion acceleration have been clearly identified by using a unique target configuration consisting of a thin CD_2 foil. Preliminary results show that protons and deuterons temperatures show opposite trend, suggesting a complex interplay between surface and volume acceleration [43] .

More experimental campaign [44] aiming at investigating the ion acceleration in laser-driven plasma with the production of a quasi-monoenergetic beam has occurred. At the LENS (Laser Energy for Nuclear Science) laboratory of INFN-LNS in Catania, experimental measures were carried out; the features of LENS are: Q-switched Nd:YAG laser with 2 J laser energy, 1064 nm fundamental wavelengths, and 6 ns pulse duration. Measures demonstrate that at the reached values of fluence even when using ns lasers, quasi-mono-energetic protons and aluminum ion beams can be generated. The presented results [45] represent a very stable and reproducible phenomenon. The production of mono-energetic protons from aluminum target was predicted by our previous work that correlated the spectrum to the initial proton distribution on the target. Preliminary results show the production of multi-charged ions having very narrow energy spreads [45-46]. The presented results are no singular manifestations, but represent a very stable and reproducible phenomenon. The production of quasi mono-energetic Al^{n} from aluminum target was predicted by our previous work [47] that correlated the spectrum to the initial ions distribution on the target. A linear dependence between the energy of the laser and the energy of the mono-peaks of ions produced is evident.

Also, we investigate the effects of innovative nanostructured targets [48] based on Ag nanowires on laser energy absorption in the ns time domain. The tested targets were realized at INFN-Bologna by anodizing aluminium sheets in order to obtain layers of porous $Al^{2}O^{3}$ of different thicknesses, on which nanowires of various metals are grown by electro-deposition with different heights. Targets were then irradiated by using a Nd:YAG laser at different pumping energies. Advanced diagnostic tools were used for characterizing the plasma plume and ion production. As compared with targets of pure Al, a huge enhancement (of almost two order of magnitude) of the X-ray flux emitted by the plasma has been observed when using the nanostructured targets, with a corresponding decrease of the "optical range" signal, pointing out that the energetic content of the laser produced plasma was remarkably increased. Signs of plasma stagnation, of great interest for carrying out inner-plasma studies, have been found [48]. Special targets, based on Co, Ni and Fe NWs, with high absorbance in the VIS and NIR range, have been irradiated with the same laser LENS. From preliminary results [49], a higher X-ray intensity was observed for all samples compared to bulk-Al, with about a ten-fold increase for Co NWs and around a five-fold increase for Ni and Fe NWs. Further tests are needed to explain such a difference among metals, which cannot be accounted in terms of different atomic numbers. Moreover, there are strong indications that plasma conditions for NWs targets are quite different from those obtained for bulk targets. This indication could be very promising, since it could open the possibility to study specific nuclear reactions in plasma, at high rates. Concerning the results on the plasma temperature analysis, NWs and the bulk-Al target seams have close temperatures. This scenario is consistent with the indication that plasma conditions for NWs targets are quite different from those obtained for bulk targets, at least in terms of plasma density and/or plasma stagnation [14]. If confirmed, these results could be very useful for nuclear physics studies, because can contribute to the increasing of the total reaction events.

All these experimental observations, although in a preliminary stage of analysis, provide a quite complete characterisation of the laser–target interaction regime in our experimental conditions and a reference set of data for the upcoming ELI-NP facility installation.

5. References

[1] Rolfs C E et al. 1988 *Cauldrons in the cosmos: nuclear astrophysics. (Chicago: University of Chicago Press. Harvard 18th ed.)*

IOP Conf. Series: Journal of Physics: Conf. Series **1014** (2018) 012006 doi:10.1088/1742-6596/1014/1/012006

[2] Negoita F et al. 2016 *Romanian Reports in Physics* **68** S37
[3] Zamfir N V 2014 *N. Eur. Phys. J. Spec. Top.* **6** 223
[4] Hoyle F 1946 *Monthly Notices of the Royal Astronomical Society* **106** 343
[5] Hoyle F. 1954 *The Astrophysical Journal Supplement Series* **1** 121
[6] Salpeter E E 1954 *Aust. J. Phys.* **7(3)** 373
[7] Angulo C 1999 *Nucl. Phys. A* **656** 3
[8] Lipoglavsˇek M 2010 *Eur. Phys. J. A* **44** 71
[9] Spitaleri C *et al.* 2011 *Phys. Atom. Nucl.* **74** 1725
[10] Bahcall N A 1969 *Fowler Astrophysical Journal* **157** 645
[11] Limata B N 2006 *Eur. Phys. J. A* ***27s01*** *193*
[12] Hannachi F et al. 2007 *Plasma Phys. Control. Fusion* ***49*** *B79*
[13] Tsallis C 2009 *Introduction to Nonextensive Statistical Mechanics (New York: Springer-Verlag)*
[14] Mascali D et al. 2010 *Rad. Eff. and Def. in Sol.* **165** 730
[15] Efimov V 2009 *Nature Phys.* **5** 533
[16] Meyer B 1994 *Annu. Rev. Astron. Astrophys.* **32** 153
[17] Coc A *et al.* 2010 *Astrophysical Journal* **744** 158
[18] Labaune C *et al.* 2013 *Nature Comm.* **4** 2506
[19] Macchi A et al. 2013 *Reviews of Modern Physics* ***85*** *751*
[20] Macchi A 2013 *A Superintense Laser-Plasma Interaction Theory Primer (Netherlands: SpringerBriefs in Physics)*
[21] Passoni M *et al.* 2008 *Phys. Rev. Lett.* **101** 115001
[22] Borghesi M *et al.* 2004 *Phys. Rev. Lett.* **92** 055003
[23] Lanzalone G *et al.* 2016 *EPJ Web of Conferences* **117** 05008
[24] Tudisco S *et al.* 2015 *Rev. Sci. Instrum.* **87** 02A909
[25] Gizzi L A *et al.* 2011 *Phys.Review ST Acc. Beams* **14** 011301
[26] Privitera S *et al.* 2008 *Sensors* **8(8)** 4636
[27] Muoio A *et al.* 2016 *EPJ Web of Conferences* **117** 10006
[28] Moll M 2003 *NIM in Physics Research A* **511** 97
[29] Sandeep K et al. 2013 *NIM A* **728** 97
[30] Gizzi L A *et al.* 2017 *Appl. Sci.* **7(10)** 984
[31] Clark E L *et al.* 2000 *Phys. Rev. Lett.* **85** 1654
[32] Hegelich O M *et al.* 2002 *Phys. Rev. Lett.* **85** 085002
[33] Fuchs J *et al.* 2006 *Nature Physics* **2** 48
[34] Passoni M *et al.* 2010 *New J. Phys.* **12** 045012
[35] Esirkepov T *et al.* 2004 *Phys. Rev. Lett.* **92** 175003
[36] Robinson A P L *et al.* 2008 *New J. Phys.* **10** 013021
[37] Robinson A P L *et al.* 2009 *Plasma Phys. Control. Fusion* **51** 024004
[38] Macchi A *et al.* 2010 *New J. Phys.* **12** 045013
[39] Qiao B *et al.* 2012 *Phys. Rev. Lett.* **108** 115002*; Kar S. et al. 2012 ibid.* **109** 185006
[40] Yin L *et al.* 2006 *Laser Part. Beams* **24** 291
[41] Yin L *et al.* 2011 *Phys. Rev. Lett.* **107** 045003
[42] Gizzi L A *et al.* 2011 *Phys. Rev. ST Accel. Beams* **14** 011301
[43] Tudisco S *et al.* 2016 *Review of Scientific Instruments* **87** 02A909
[44] Altana C *et al.* 2016 *Review of Scientific Instruments* **87** 02A914
[45] Muoio A *et al.* 2016 *Nuclear Instruments and Methods in Physics Research A* **829** 141
[46] Altana C *et al.* 2016 *Nuclear Instruments and Methods in Physics Research A* **829** 159
[47] Mascali D *et al.* 2012 *EPL(Europhysics Letters)* **100** 45003
[48] Lanzalone G *et al.* 2016 *Review of Scientific Instruments* **87** 02B324
[49] Muoio A *et al.* 2017 *Journal of Instrumentation* **12** C03076

IOP Conf. Series: Journal of Physics: Conf. Series **1014** (2018) 012007 doi:10.1088/1742-6596/1014/1/012007

Entrance channel effects on the deexcitation ways of the same compound nucleus at a fixed excitation energy

A. Anastasi[1], G. Mandaglio[2,3], F. Curciarello[4], A. K. Nasirov[5,6], G. Fazio[1], G. Giardina[1]

[1] Dipartimento di Scienze Matematiche e Informatiche, Scienze Fisiche e Scienze della Terra, University of Messina, Messina, Italy
[2] Dipartimento di Scienze Chimiche, Biologiche, Farmaceutiche ed Ambientali, University of Messina, Messina, Italy
[3] INFN Sezione di Catania, Catania, Italy
[4] INFN Laboratori Nazionali di Frascati, Frascati, Italy
[5] JINR - Bogoliubov Laboratory of Theoretical Physics, Dubna, Russia
[6] National University of Uzbekistan, Tashkent, Uzbekistan

E-mail: aanastasi@unime.it, gmandaglio@unime.it

Abstract. The investigation of various properties of deexcitation of the same ^{220}Th compound nucleus (CN), formed by the different mass (charge) asymmetric ^{16}O+^{204}Pb, ^{40}Ar+^{180}Hf, ^{82}Se+^{138}Ba and ^{96}Zr+^{124}Sn reactions is presented. The effective fission barrier $< B_{\rm fis} >$ value, as a function of the excitation energy $E^*_{\rm CN}$, determined for each intermediate excited nucleus reached along the deexcitation cascade of the CN obtained by the four considered reactions is strongly sensitive to the various orbital angular momentum L= $\ell\hbar$ distributions of CN formed with the same excitation energy E^*_{CN} by the various entrance channels. Therefore, the competition between the fission and evaporation of light particles (neutron, proton, and α-particle) processes along the deexcitation cascade of CN is dependent on the orbital angular momentum distribution of CN. In fact, the ratio between the evaporation residue cross sections obtained when also the charged particles are emitted and the ones obtained after neutron emission only for the same CN with a fixed excitation energy E^*_{CN} is sensitive to the mass (charge) asymmetry of the entrance channel.

1. Introduction

It is well known to the scientific community that in heavy ion collisions, at low energies, the complexity of processes preceding the formation of reaction products strongly influences the properties and nature of these final products, and that due to very transient characteristics of these processes it is impossible to observe how they occur. Moreover, in the reactions there are products that are strongly determined by the first stage of the collision between the projectile and the target nuclei leading to the capture of reactants and then to the evolution of the dinuclear system (DNS) [1] up to the formation of products of the quasifission process in competition with the ones of the complete fusion process. In this last case, the complete fusion stage can lead to the fast fission products (for angular momentum values $\ell > \ell_{\rm cr}$) and reaction products determined by the deexcitation of the compound nucleus (CN) that leads to the fission fragments in competition with the evaporation residue (ER) nuclei after light particle emissions surviving fission [2–6]. In this complex context, many quasifission, fast fission and fusion-fission products are overlapped, and many ER nuclei can not be detected and identified to cause of concrete limits of experimental apparata and/or analysis of data. Therefore, in the analysis of experimental data there are unavoidable uncertainties on the identification and separation of the products that are formed in each step of the reaction process. Of course, also in calculation of the theoretical models there are serious uncertainties on the obtained results due to the assumptions made in the procedures and use of phenomenological models. In this paper, we present a detailed analysis and comparison of calculated results obtained by the study of the ^{16}O+^{204}Pb, ^{40}Ar+^{180}Hf, ^{82}Se+^{138}Ba and ^{96}Zr+^{124}Sn reactions leading to the same ^{220}Th CN. Since the products of reactions at deexcitation of CN are the evaporation residues after neutrons and charged particle emissions, and the fission fragments produced in competition with the light particle emission along the various steps of the deexcitation cascade of CN, it is possible to analyze the effects related to the various modes of deexcitation of the same formed CN with the same excitation energy $E^*_{\rm CN}$ but characterized by a different angular momentum distribution with respect to the various mass (charge) asymmetry in the entrance channel.

2. Method

The study of heavy ion collisions near the Coulomb barrier energies is based on calculations of the incoming path of projectile nucleus and finding the capture probability, taking into account the possibility of interaction with different orientation angles of the axial symmetry axis of deformed nuclei [4]. Also, the surface vibration of the nuclei, which are spherical in the ground state and deformed shape in the first excited 2^+ state, is taken into consideration. The final results are averaged over all orientation angles the axial symmetry axis of deformed nuclei or vibrational states of the spherical nuclei. These procedures are presented in the Appendix A and Appendix B of the paper [2].

The capture of the projectile by the target is characterized by the full momentum transfer of the relative momentum into the intrinsic degrees of freedom and shape deformation. The capture occurs if the following necessary and sufficient conditions are satisfied. The necessary condition of capture is overcoming the Coulomb barrier by projectile nucleus to be trapped in the potential well of the potential energy surface (PES). The collision dynamics is calculated by the solution of the equations of the relative distance R and angular momentum L [2].

The condition of sufficiency for capture is the decrease of the relative kinetic energy due to dissipation by friction forces up to values lower than the depth of the potential well [7–9]. The potential well is formed due to the competition of the short range nuclear attractive and the Coulomb and centrifugal repulsive potentials. This condition depends on the values of the beam energy and orbital angular momentum, the size of the potential well and intensity of the friction forces that cause dissipation of the kinetic energy of the relative motions to internal energy of two nuclei. So, the trapping of the collision path in the well means that the capture has occurred and the DNS is formed. The lifetime of the DNS is determined by its excitation energy $E^*_{\rm DNS}$ and by the size of the potential well. The height of the inner barrier of the potential well is called the quasifission barrier $B_{\rm qfis}$ in our approach. This definition is related to the quasifission process: in this case the DNS decays without reaching the equilibrated shape of a compound nucleus [8–10]. The alternative to the quasifission process in the evolution of DNS is the complete fusion of its constituent fragments. According to this scenario the partial capture cross section $\sigma^\ell_{\rm cap}$ for a given relative energy in the center-of-mass system $E_{\rm c.m.}$ and angular momentum value ℓ is the sum of the partial complete fusion $\sigma^\ell_{\rm fus}$ and quasifission $\sigma^\ell_{\rm qfis}$ cross sections [11];

$$\sigma^\ell_{\rm cap}(E_{\rm c.m.},\ell;\alpha_1,\alpha_2) = \sigma^\ell_{\rm fus}(E_{\rm c.m.},\ell;\alpha_1,\alpha_2) + \sigma^\ell_{\rm qfis}(E_{\rm c.m.},\ell;\alpha_1,\alpha_2). \quad (1)$$

The capture cross section is determined by the number of partial waves which lead the path of the total energy of colliding nuclei to be trapped in the well of the nucleus-nucleus potential after dissipation of a sufficient part of the initial kinetic energy. The size of the potential well decreases with increasing orbital angular momentum ℓ. Therefore, the capture cross section is calculated by the formula

$$\sigma_{\text{cap}}(E_{\text{c.m.}};\alpha_1,\alpha_2) = \sum_{\ell=0}^{\ell_d(E_{\text{c.m.}})} \sigma^{\ell}_{\text{cap}}(E_{\text{c.m.}},\ell;\alpha_1,\alpha_2) = \frac{\lambda^2}{4\pi}\sum_{\ell=0}^{\ell_d(E_{\text{c.m.}})} (2\ell+1)\times \mathcal{P}^{\ell}_{\text{cap}}(E_{\text{c.m.}},\ell;\alpha_1,\alpha_2), \tag{2}$$

the fusion cross section is obtained as

$$\sigma_{\text{fus}}(E_{\text{c.m.}};\alpha_1,\alpha_2) = \sum_{\ell=0}^{\ell_d(E_{\text{c.m.}})} \sigma^{\ell}_{\text{cap}}(E_{\text{c.m.}},\ell;\alpha_1,\alpha_2)\times P^{\ell}_{CN}(E_{\text{c.m.}},\ell;\alpha_1,\alpha_2), \tag{3}$$

and the quasifission cross section is obtained as

$$\sigma_{\text{qfis}}(E_{\text{c.m.}};\alpha_1,\alpha_2) = \sum_{\ell=0}^{\ell_d(E_{\text{c.m.}})} \sigma^{\ell}_{\text{cap}}(E_{\text{c.m.}},\ell;\alpha_1,\alpha_2)\times [1 - P^{\ell}_{CN}(E_{\text{c.m.}},\ell;\alpha_1,\alpha_2)]; \tag{4}$$

for details see [3].

Here λ is the de Broglie wavelength of the entrance channel and $\mathcal{P}^{\ell}_{capture}(E_{\text{c.m.}},\ell;\alpha_1,\alpha_2)$ is the capture probability which depends on the collision dynamics: is 1 at $\ell_{\text{min}} \le \ell \le \ell_d$, while is 0 if $\ell < \ell_{\text{min}}$ or $\ell > \ell_d$.

That means the possible values leading to capture can form a "window" of angular momentum values and consequently the friction coefficient is not so strong to trap the projectile in the potential well. Moreover, the maximal value of partial waves (ℓ_d) leading to capture is calculated by the solution of the equations of the relative motion of nuclei [7,8,12], and ℓ_{min} is the minimal value of ℓ leading to capture.

3. Role of the angular momentum on the CN formation and its deexcitation

The role of the angular momentum distribution of the entrance channel on the capture and fusion cross sections and consequently on the evaporation residue products are widely discussed in our previous papers [2, 3]. Here, we like to discuss about the possibilities offered by our codes [2, 3] to calculate for each step of reaction the main properties of various processes. We obtain that:

(i) The competition between the quasifission and complete fusion at the deexcitation of DNS is strongly sensitive to the characteristics of reactants in the entrance channel and the orbital angular momentum distribution; the main role is played by the values of the B^*_{fus} intrinsic fusion barrier and B_{qfis} quasifission barrier that are both sensitive to the angular momentum ℓ values [13–15].

(ii) The contributions of the fast fission process and the compound nucleus formation are sensitive to the entrance channel through the angular momentum distribution characterizing the excitation functions of the related cross sections.

(iii) The contributions of products formed along the deexcitation cascade of CN trough the competition between the processes: evaporation of light particles (that can lead to the formation of ER nuclei with Z_i and A_i values not very different from the Z and A numbers of CN), and fission process (that lead definitively to two nuclear fragments with Z_i and A_i values near $Z/2$ and $A/2$, where Z and A are the atomic and mass number of CN, respectively).

IOP Conf. Series: Journal of Physics: Conf. Series **1014** (2018) 012007 doi:10.1088/1742-6596/1014/1/012007

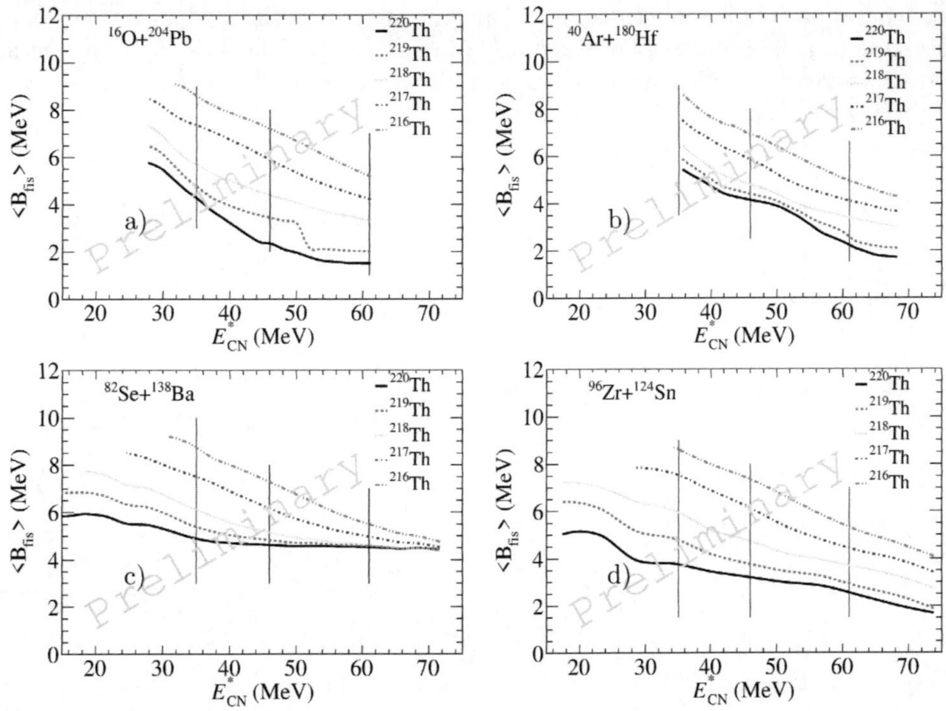

Figure 1. The effective fission barrier $< B_{\text{fis}} >$ for the excited ^{220}Th* CN and some of its isotopes ^{219,218,217,216}Th* formed after the successive neutron evaporation processes by the ^{16}O+^{204}Pb, ^{40}Ar+^{180}Hf, ^{82}Se+^{138}Ba and ^{96}Zr+^{124}Sn reactions presented in panels a), b), c) and d), respectively.

Therefore, starting from the CN we calculate for each intermediate excited nucleus formed along the deexcitation cascade the Γ_n, Γ_p, Γ_α, and Γ_{fis} widths that are dependent on the values of the separation energies of emitted particles and the fission barrier, respectively, as well as the level densities of the successive excited intermediate nuclei and the produced evaporation residues, or the obtained fission fragments. The shell corrections in the fission barriers and the level densities are damped with the excitation energies E^* of nuclei and with the orbital angular momentum ℓ. For details see [2,3] and references therein.

The method of our study for the deexcitation of CN does not use free parameters that can be changed for each reaction, but it is rigorously applied to all reactions and at any explored excited energy range.

4. Results on reactions leading to the ^{220}Th CN

We study some properties of the deexcitation of the ^{220}Th CN formed by four very different mass (charge) asymmetry and almost symmetric reactions in the entrance channel (^{16}O+^{204}Pb, ^{40}Ar+^{180}Hf, ^{82}Se+^{138}Ba and ^{96}Zr+^{124}Sn) in a wide range of excitation energy E^*_{CN}.

In Fig. 1 panel a), we present the effective fission barrier $< B_{\text{fis}} >$ versus the excitation energy E^*_{CN} obtained for the investigated ^{16}O+^{204}Pb very asymmetric reaction at first step of the deexcitation of ^{220}Th* (full line), at deexcitation of ^{219}Th* (dashed line) after 1 neutron emission from the ^{220}Th CN. Moreover, we present the excitation function of $< B_{\text{fis}} >$ for ^{218}Th* (dotted line), ^{217}Th* (dash-dotted line), and ^{216}Th* (dash-double-dotted line) after one neutron emission from ^{219}Th*, ^{218}Th*, and ^{217}Th* respectively.

IOP Conf. Series: Journal of Physics: Conf. Series **1014** (2018) 012007 doi:10.1088/1742-6596/1014/1/012007

The effective fission barrier $< B_{\rm fis} >$ value of a intermediate excited nucleus is obtained as the weighted average on the partial production cross section $\sigma^{\ell}_{\rm fus}$ [16] of all possible $B_{\rm fis}(\ell, T)$ values defined as:

$$B_{\rm fis}(\ell, T) = cB^{\rm m}_{\rm fis} - h(T)q(\ell)\delta W \tag{5}$$

where c=1, while $h(T)$ and $q(\ell)$ represent the damping functions of the nuclear shell correction δW by the increase of the excitation energy E^* and angular momentum ℓ, respectively. The nuclear temperature T is defined as $\sqrt{E^*/a}$, where a is the intrinsic level density parameter [3, 17].

Analogously, we present in Fig. 1 panels b), c), and d) the excitation functions of $< B_{\rm fis} >$ for the ^{40}Ar+^{180}Hf asymmetric reaction, ^{82}Se+^{138}Ba almost symmetric reaction, and ^{96}Zr+^{124}Sn symmetric reaction, respectively.

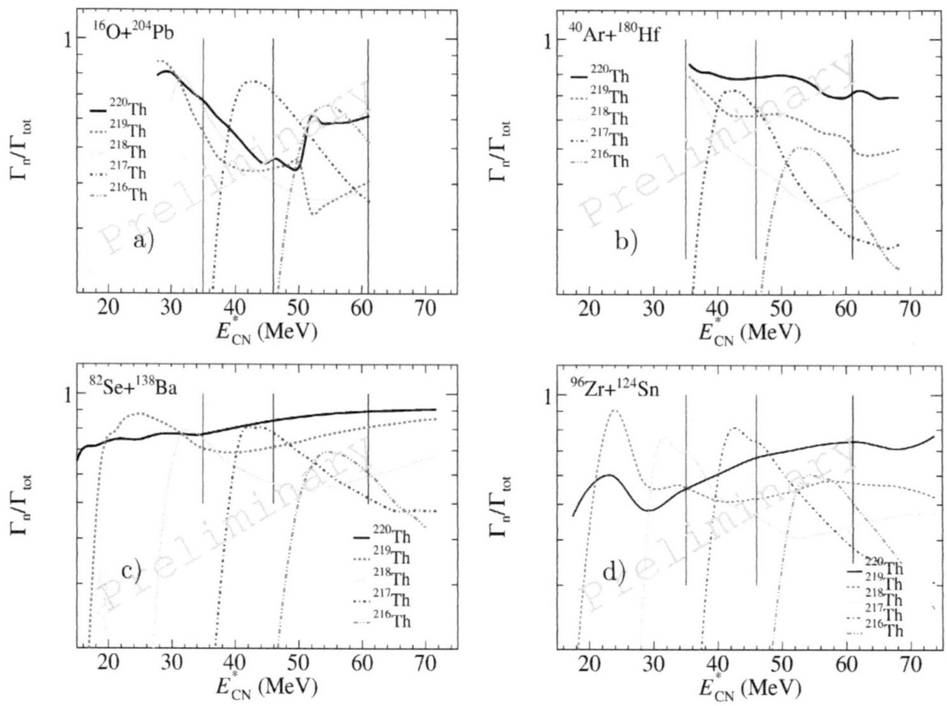

Figure 2. The excitation functions of $\Gamma_{\rm n}/\Gamma_{\rm tot}$ at the deexcitation of the ^{220}Th* CN and the ^{219,218,217,216}Th* intermediate excited nuclei for the reactions and panels as indicated in Fig. 1.

In panels a), b), c), and d) of Fig. 2 are presented the calculated results of excitation functions $\Gamma_{\rm n}/\Gamma_{\rm tot}$ neutron emission probability from the ^{220}Th*, ^{219}Th*, ^{218}Th*, ^{217}Th*, and ^{216}Th* excited nuclei for the ^{16}O+^{204}Pb, ^{40}Ar+^{180}Hf, ^{82}Se+^{138}Ba and ^{96}Zr+^{124}Sn very different entrance channels, respectively.

In panels a), b), c), and d) of Fig. 3 are reported the results of $\Gamma_{\rm fis}/\Gamma_{\rm tot}$ excitation functions representing the fission probability from the ^{220}Th*, ^{219}Th*, ^{218}Th*, ^{217}Th*, and ^{216}Th* excited nuclei for the above-mentioned reactions.

By observing the values and trends of the $< B_{\rm fis} >$, $\Gamma_{\rm n}/\Gamma_{\rm tot}$, and $\Gamma_{\rm fis}/\Gamma_{\rm tot}$ excitation functions versus $E^*_{\rm CN}$ reported in Figs. 1, 2, and 3, respectively, for the four investigated reactions leading to the same ^{220}Th* CN, it is possible to ascertain the variations of single values and the complete interval of the obtained excitation function for the ^{220}Th*, ^{219}Th*, ^{218}Th*, ^{217}Th*, and ^{216}Th* excited nuclei along the five steps of the deexcitation cascade of CN. In order to have a clearer

IOP Conf. Series: Journal of Physics: Conf. Series **1014** (2018) 012007 doi:10.1088/1742-6596/1014/1/012007

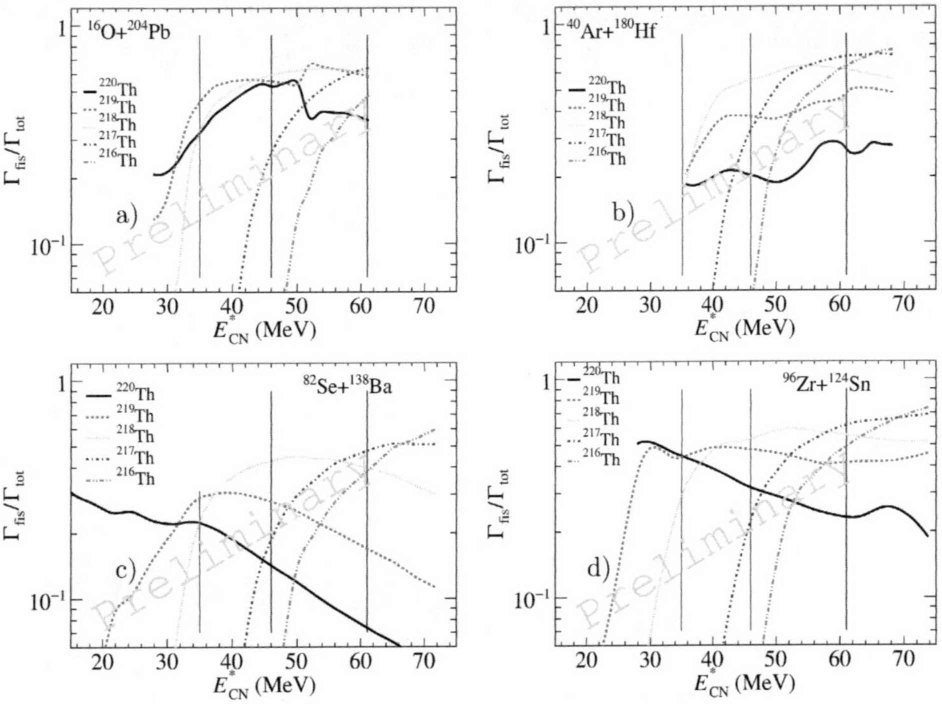

Figure 3. As Fig. 2 but for the fission probability $\Gamma_{fis}/\Gamma_{tot}$.

observation and comparison of results, we choose and indicate in figures by vertical lines the values of 35, 46, and 61 MeV of excitation energy of CN and we compare the obtained values of the studied functions related to the considered excited nuclei. It is easy to see that the values found for the various excited nuclei ^{220,219,218,217,2016}Th* are significantly different for the four reactions studied (and presented in panels a), b), c) and d), respectively) at 35, 46 and 61 MeV. In fact, in Fig. 1 where are reported the values of the effective fission barrier $< B_{fis} >$ versus E^*_{CN}, in panel a) related to ^{16}O+^{204}Pb reaction, at E^*_{CN} = 35 MeV $< B_{fis} >$ for the ^{220}Th* is 4.3 MeV while for the last studied step of ^{216}Th* the value of $< B_{fis} >$ is 8.6 MeV, in panel b) related to the ^{40}Ar+^{180}Hf reaction, $< B_{fis} >$ for ^{220}Th* is 5.5 MeV while for ^{216}Th* is 8.8 MeV; analogously,in panel c) for the ^{82}Se+^{138}Ba reaction $< B_{fis} >$ for the ^{220}Th* CN is 4.9 MeV while for ^{216}Th* is 8.73 MeV, and in panel d) for the ^{96}Zr+^{124}Sn reaction, $< B_{fis} >$ for the ^{220}Th* CN is 3.7 MeV while for ^{216}Th* is 8.6 MeV. It is possible to verify that also for other intermediate excited nuclei ^{219,218,217}Th* the corresponding $< B_{fis} >$ are generally different for the considered reactions at the same excitation energy E^*_{CN}. These results are determined by the different angular momentum distributions of the partial fusion cross section for the four different entrance channels, even when the excitation energy E^*_{CN} is the same (see figure 4 at E^*_{CN} = 35, 46 and 61 MeV for the excitation energies of the formed CN by the four considered reactions in the entrance channel).

An analogous behavior is ascertained at E^*_{CN}=46 MeV : in panel a), for the reaction induced by ^{16}O, the $< B_{fis} >$ value for ^{220}Th* is 2.35 MeV while for ^{216}Th* is 7.2 MeV; in panel b), for the reaction induced by ^{40}Ar, $< B_{fis} >$ for the ^{220}Th* CN is 4.1 MeV while for ^{216}Th* is 6.9 MeV; in panel c), for the reaction induced by ^{82}Se, $< B_{fis} >$ for the ^{220}Th* CN is 4.6 MeV while for ^{216}Th* $< B_{fis} >$ is 7.3 MeV; in panel d), for the reaction induced by ^{96}Zr, $< B_{fis} >$ for the ^{220}Th* CN is 3.2 MeV while for ^{216}Th* $< B_{fis} >$ is 7.3 MeV.

IOP Conf. Series: Journal of Physics: Conf. Series **1014** (2018) 012007 doi:10.1088/1742-6596/1014/1/012007

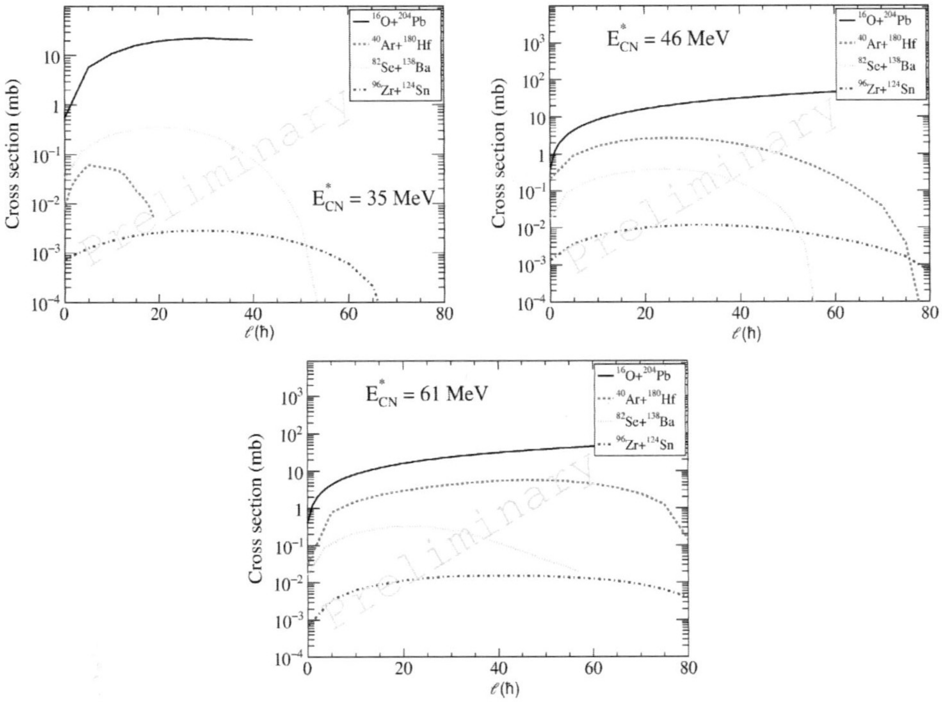

Figure 4. Partial fusion cross sections $\sigma^{\ell}_{\rm fus}$ as a function of angular momentum ℓ for three fixed excitation energies of $E^*_{\rm CN}$=35, 46, and 61 MeV: see insert for details.

Finally, at $E^*_{\rm CN}$=61 MeV we obtain: in panel a), $< B_{\rm fis} >$ for ^{220}Th* is 1.5 MeV while for ^{216}Th* $< B_{\rm fis} >$ is 5.2 MeV; in panel b) $< B_{\rm fis} >$ for the ^{220}Th* CN is 2.25 MeV while for ^{216}Th* is 5.0 MeV; in panel c), $< B_{\rm fis} >$ for the ^{220}Th* CN is 4.5 MeV while for ^{216}Th* $< B_{\rm fis} >$ is 5.5 MeV; in panel d), $< B_{\rm fis} >$ for the ^{220}Th* CN is 2.6 MeV while for ^{216}Th* $< B_{\rm fis} >$ is 5.3 MeV.

Similar considerations can be made by observing the values of neutron emission probabilities $\Gamma_n/\Gamma_{\rm tot}$ reported in Fig. 2 and the values for the fission probabilities $\Gamma_{\rm fis}/\Gamma_{\rm tot}$ reported in Fig. 3, at any fixed excitation energy value of $E^*_{\rm CN}$ when the ^{220}Th CN is formed by the four very different reaction in the entrance channel. Since the effective fission barrier $< B_{\rm fis} >$ values are significantly different for the excited nuclei ^{220}Th*, ^{219}Th*, ... ^{216}Th* along the deexcitation cascade at any fixed excitation energy value of the formed ^{220}Th CN, also the competition between the neutron emission probability $\Gamma_n/\Gamma_{\rm tot}$ value and the fission probability $\Gamma_{\rm fis}/\Gamma_{\rm tot}$ value is different for any excited nucleus reached along the deexcitation cascade, even when the excitation energy $E^*_{\rm CN}$ of the ^{220}Th CN is the same but the mass (charge) asymmetry of reaction in the entrance channel is different. For this reason, we also report in all panels a), b), c) and d) of Fig. 2 and Fig. 3 , three vertical lines at 35, 46 and 61 MeV values of excitation energy of the ^{220}Th in order to easier observe the variations of $\Gamma_n/\Gamma_{\rm tot}$ and $\Gamma_{\rm fis}/\Gamma_{\rm tot}$ values for the same reached excited nucleus ^{220}Th*, ^{219}Th*, ... ^{216}Th* when any individuated excited nucleus is formed by the four above-mentioned different entrance channels.

5. Conclusion

The present study on heavy ion reactions with various mass (charge) asymmetry parameters points out the effects of the entrance channel on the CN formation and the consequent different ways of its deexcitation cascade even when the formed CN is characterized by the same Z and A values and has the same excitation energy $E^*_{\rm CN}$. The reason of this different way of deexcitation is due to the different orbital angular momentum distribution of reactants in the entrance channel to cause of different mass (charge) asymmetry parameter of reacting nuclei, of the specific shapes (oblate, prolate or spherical), and also of the eventual deformation parameters of these beam and target nuclei, even when the CN is formed with the same Z, A and $E^*_{\rm CN}$ values. Moreover, the effective fission barrier $< B_{\rm fis} >$ of each intermediate excited nucleus reached at each step of the deexcitation cascade of the same formed CN with the same $E^*_{\rm CN}$ but by various entrance channels, is affected by the various angular momentum distributions for the four considered reactions forming the ^{220}Th CN; moreover, the damping function of the fission barrier $B_{\rm fis}$ determines different effects on the deexcitation of CN to cause of the different range of the angular momentum ℓ for the considered reactions. Therefore, the competition between $\Gamma_{\rm n}/\Gamma_{\rm tot}$ and $\Gamma_{\rm fis}/\Gamma_{\rm tot}$ for each considered intermediate excited nucleus reached along the deexcitation cascade of CN is affected by the type of reaction in the entrance channel and the excitation energy $E^*_{\rm CN}$ too. In addition, we can anticipate the information that at each $E^*_{\rm CN}$ value the formation of evaporation residue nuclei ERs –formed after neutrons emission only along the deexcitation cascade of CN and the ones formed by the charged emission too– is strongly sensitive to the type of reaction in the entrance channel and to the complete reaction mechanism that is specific of each reaction also when it leads to the same CN for A, Z and $E^*_{\rm CN}$ values.

Therefore, by considering also the emission of charged particles proton and α together with neutron emission along the deexcitation cascade of the ^{220}Th CN the results of calculation show that the total ER evaporation residue nuclei cross sections contributed by the intermediate excited nuclei after charged particle emission too (with Z <90) along the complete deexcitation cascade of CN are greater than the ER yields obtained by the excited nuclei after neutron emission only (with Z =90) [2]. For example, at $E^*_{\rm CN} = 46$ MeV, the ratio ER($Z < 90$)/ER($Z =$ 90) values are 8.2, 6.0, 10.1, and 8.7 for the reaction induced by the ^{16}O, ^{40}Ar, ^{82}Se, and ^{96}Zr, respectively. This additional preliminary result on the evaporation residue production furtherly confirms that the deexcitation of the same CN with the same excitation energy $E^*_{\rm CN}$ formed by various reactions with different mass (charge) asymmetry parameter is strongly sensitive to the effects of the entrance channel and to the complete reaction mechanism that is specific for each reaction.

By our procedure it is also possible to determine the excitation function of the neutron emission multiplicities $\nu_{\rm pre}$ overcoming the fission process as a function of the excitation energy $E^*_{\rm CN}$, and we also observe for the $\nu_{\rm pre}$ determination the effect of the entrance channel on the related excitation functions.

References

[1] Volkov V V *et al.* 1993 *Phys. Lett. B* **319** 425, 1995 *Phys. Rev. C* **51** 2635
[2] Kim K *et al.* 2015 *Phys. Rev. C* **91** 064608
[3] Giardina G *et al.* 2018 *Nucl. Phys. A* **970** 169
[4] Fazio G *et al.* 2008 *J. Phys. Soc. Jpn.* **77** 124201
[5] Mandaglio G *et al.* 2009 *Phys. At. Nucl.* **72** 1639
[6] Mandaglio G *et al.* 2012 *Phys. Rev. C* **86** 064607
[7] Fazio G *et al.* 2005 *Phys. Rev. C* **72** 064614
[8] Nasirov A K *et al.* 2005 *Nucl. Phys. A* **759** 342
[9] Fazio G *et al.* 2005 *Mod. Phys. Lett. A* **20** 391
[10] Back B B *et al.* 1985 *Phys. Rev. C* **32** 195
[11] Nasirov A K *et al.* 2010 *Phys. Lett. B* **686** 72
[12] Fazio G *et al.* 2003 *J. Phys. Soc. Jpn.* **72** 2509
[13] Antonenko N V *et al.* 1993 *Phys. Lett. B* **319** 425
[14] Antonenko N V *et al.* 1995 *Phys. Rev. C* **51** 2635
[15] Fazio G *et al.* 2004 *Eur. Phys. J. A* **22** 75

IOP Conf. Series: Journal of Physics: Conf. Series **1014** (2018) 012007 doi:10.1088/1742-6596/1014/1/012007

[16] Anastasi A *et al.* 2015 *Acta Phys. Pol. B* **8** 583
[17] Ignatyuk A V *et al.* 1975 *Sov. J. Nucl. Phys.* **21** 255

IOP Conf. Series: Journal of Physics: Conf. Series **1014** (2018) 012008 doi:10.1088/1742-6596/1014/1/012008

Cluster formation in nuclear reactions from mean-field inhomogeneities

Paolo Napolitani[1], Maria Colonna[2] and Carlo Mancini-Terracciano[3]

[1] IPN, CNRS/IN2P3, Université Paris-Sud, Université Paris-Saclay, 91406 Orsay, France
[2] INFN-LNS, Laboratori Nazionali del Sud, 95123 Catania, Italy
[3] INFN, Sezione di Roma, Rome, Italy

Abstract. Perturbing fluids of neutrons and protons (nuclear matter) may lead, as the most catastrophic effect, to the rearrangement of the fluid into clusters of nucleons. A similar process may occur in a single atomic nucleus undergoing a violent perturbation, like in heavy-ion collisions tracked in particle accelerators at around 30 to 50 MeV per nucleon: in this conditions, after the initial collision shock, the nucleus expands and then clusterises into several smaller nuclear fragments.

Microscopically, when violent perturbation are applied to nuclear matter, a process of clusterisation arises from the combination of several fluctuation modes of large-amplitude where neutrons and protons may oscillate in phase or out of phase. The imposed perturbation leads to conditions of instability, the wavelengths which are the most amplified have sizes comparable to small atomic nuclei. We found that these conditions, explored in heavy-ion collisions, correspond to the splitting of a nucleus into fragments ranging from Oxygen to Neon in a time interval shorter than one zeptosecond (10^{-21}s). From the out-of-phase oscillations of neutrons and protons another property arises, the smaller fragments belonging to a more volatile phase get more neutron enriched: in the heavy-ion collision case this process, called distillation, reflects in the isotopic distributions of the fragments.

The resulting dynamical description of heavy-ion collisions is an improvement with respect to more usual statistical approaches, based on the equilibrium assumption. It allows in fact to characterise also the very fast early stages of the collision process which are out of equilibrium. Such dynamical description is the core of the Boltzmann-Langevin One Body (BLOB) model, which in its latest development unifies in a common approach the description of fluctuations in nuclear matter, and a predictive description of the disintegration of nuclei into nuclear fragments. After a theoretical introduction, a few practical examples will be illustrated.

This paper resumes the extended analysis of fluctuations in nuclear matter of ref. [2] and briefly reviews applications to heavy-ion collisions.

1. Introduction

The most catastrophic process which can occur in a nuclear complex is its splitting into clusters and fragments when undergoing a violent external action. We want to address this process, which can be probed in a dissipative heavy-ion collision, from the point of view of dynamics, moving from nuclear matter to nuclei, which are finite open self-bound systems. Nuclear clusterisation may appear in various forms. Exotic topologies in nuclear astrophysics (stellar matter), nuclear states explored in low-energy reactions from the recombination and vibrations of existing cluster structures, low-density regions of the equation of state landscape, where clusterisation in intermediate-mass fragments and clusters may arise and can be explained in violent nuclear reactions. Finally, clusterisation characterises in general Fermi liquids as arising from ripples produced by phase-space fluctuations [1].

We constructed a microscopic dynamical framework from applying the theory of Fermi liquids to clusterisation in nuclei (see ref. [2] for a more extended and detailed discussion); within this framework, we explore how the clusterisation progresses from zero-sound propagation. Two avenues are mostly followed to describe nuclear processes, molecular dynamics and mean-field approaches [3]. In mean-field approaches particles evolve independently in their own self-consistent mean-field. Such treatment, while well suited to describe the collective behaviour, can not handle clusterisation in its pure mean-field form. On the other hand, molecular-dynamics approaches rely on a description in terms of product wave functions. The independent-particle scheme (i.e. mean field) is applied to nucleonic single-particle wavefunctions ψ_i, so that many-body correlations in both mean field and scattering can be achieved from the localisation of ψ_i (a coherent-state subspace is used in FMD [6] and an even stronger localisation is imposed in AMD [4, 5] by also fixing the widths of ψ_i). Such treatment is successful for final-state correlations, but it may approximate collective behaviour and 0-sound propagation.

To profit from a well suited description of 0-sound propagation, we follow thereafter a mean-field approach where, in order to describe clusterisation, extensions to handle large-amplitude dynamics should be explicitly introduced. In particular, correlations beyond the level of kinetic equations are needed or, in terms of BBGKY hierarchy [7], upper orders beyond two-body correlations should be recovered in order to access highly non-linear regimes. A technique to obtain such result in an approximated form consists in applying two treatments in parallel: first, introducing nucleon-nucleon correlations continuously, second, handling a stochastic ensemble of several mean-field trajectories to exploit the introduced correlations. In this case, a mean-field trajectory ρ_1 taken at a given time, would split at a successive time into subensembles $\rho_1^{(n)}$:

$$\rho_1 \longrightarrow \{\rho_1^{(n)}; n = 1, \ldots, \text{subens.}\} \ . \tag{1}$$

In practice, at the level of kinetic equations, not only collisional correlations should be introduced, but also a term of vanishing mean which injects fluctuations around the collision integral intermittently in time. Such term can be exploited as a stochastic source to revive fluctuations all along a dissipative process. Adding the collisional correlations introduced above and fluctuations to the mean-field produces a scheme which resembles stochastic TDHF [8, 9]:

$$i\hbar \frac{\partial \rho_1^{(n)}}{\partial t} \approx [k_1^{(n)} + V_1^{(n)}, \rho_1^{(n)}] + \bar{I}_{\text{coll}}^{(n)} + \delta I_{\text{coll}}^{(n)} \ , \tag{2}$$

where $\bar{I}_{\text{coll}}^{(n)}$ and $\delta I_{\text{coll}}^{(n)}$ are the average collision contribution and a continuous source of fluctuation seeds, respectively. The corresponding Wigner transform yields the Boltzmann-Langevin (BL) equation [10], in terms of an ensemble of distribution function f^n,

$$\frac{\partial f^{(n)}}{\partial t} = \{h^{(n)}, f^{(n)}\} + I_{\text{UU}}^{(n)} + \delta I_{\text{UU}}^{(n)} \ , \tag{3}$$

where the ensemble $f^{(n)}$ replaces the ensemble of Slater determinants in Eq. (2) and corresponds to a Fermi statistics at equilibrium, $h^{(n)}$ is the effective Hamiltonian acting on $f^{(n)}$, and the residual contributions of Eq. (2) are replaced by corresponding Uehling-Uhlenbeck (UU) terms.

We focus on the BL equation and we solve it in full phase space through the BLOB approach [11, 12, 2], where nucleon-nucleon correlations are introduced by constructing a fluctuating collision term which acts on extended equal-isospin phase-space portions for each single in-medium collision:

$$\frac{\partial f^{(n)}}{\partial t} - \{h^{(n)}, f^{(n)}\} = I_{\text{UU}}^{(n)} + \delta I_{\text{UU}}^{(n)} = g \int \frac{\mathrm{d}\mathbf{p}_b}{h^3} \int W(\text{AB}{\leftrightarrow}\text{CD})\, F(\text{AB}{\rightarrow}\text{CD})\, \mathrm{d}\Omega \ , \tag{4}$$

IOP Conf. Series: Journal of Physics: Conf. Series **1014** (2018) 012008 doi:10.1088/1742-6596/1014/1/012008

where g is the degeneracy factor, W is the transition rate in terms of relative velocity between the two colliding phase-space portions, and F handles the Pauli blocking of initial and final states over their full phase-space extensions. The extended size of phase-space portions involved in the scattering in Eq. (4) should be large enough so that the occupancy variance in h^3 cells is equal to $f(1-f)$, which corresponds to the scattering of two nucleons [13]. For comparison, we consider a second simplified approach to solve the BL equation, Eq. (3), based on the SMF [14] treatment, where fluctuations are injected from an external stochastic contribution U_{ext} and projected on spacial density.

In both cases, in all following calculations, a simplified SKM* effective interaction, with momentum dependence omitted, is used [15, 16]. A soft isoscalar equation of state with a compressibility modulus $k = 200\text{MeV}$ is used. For test, both asy-stiff (linear) and asy-soft (quadratic) parametrisations are used for the symmetry energy. All nuclear-matter calculations employ a periodic box of edge $L = 39\text{fm}$, and the system is initialised with a Fermi-Dirac distribution at a temperature $T = 3\text{MeV}$. In the collision term, either a free or a constant nucleon-nucleon cross section σ_{NN} is used.

Before carrying on a study on heavy-ion collisions, we require that fluctuation amplitudes are consistent with analytic expectations from Fermi liquids: for this purpose, we study nuclear matter in initially homogeneous conditions.

2. Fluctuations in two-component nuclear matter

While unperturbed, the dynamics of a periodic portion of uniform nuclear matter at low temperature (so that the collision term I_{UU} in the analytic description can be neglected) would evolve along a mean trajectory f^0. Let us introduce perturbations $\delta f^q \ll f^0$ in two forms: either neutrons and protons move in phase (isoscalar perturbation, indexed with q = s), so that $\delta f^{\text{q}} = \delta f^{\text{s}} = (f_{\text{n}} - f^0_{\text{n}}) + (f_{\text{p}} - f^0_{\text{p}})$, or out of phase (isovector perturbation, indexed with q = v), so that $\delta f^{\text{q}} = \delta f^{\text{v}} = (f_{\text{n}} - f^0_{\text{n}}) - (f_{\text{p}} - f^0_{\text{p}})$. In a periodic box, this action introduces fluctuations ρ^{q}_k associated with plane waves of wave number $\mathbf{k}$, with a corresponding equilibrium variance $(\sigma^{\text{q}}_k)^2$ (or intensity of response), and with an equilibrium variance of spacial density correlations $(\sigma_{\rho^{\text{q}}})^2$ (this latter variance is obtained from the former through an inverse Fourier transform).

If the Boltzmann-Langevin equation is applied to the disturbance δf^{q}, the following scheme can be established.

- First, if fluctuations are of stable nature, the two equilibrium variances $(\sigma^{\text{q}}_k)^2$ and $(\sigma_{\rho^{\text{q}}})^2$ can be related to the free-energy density curvature $F^{\text{q}}(k)$ through the fluctuation-dissipation theorem:
$$(\sigma^{\text{q}}_k)^2 = \frac{T}{F^{\text{q}}(k)} \; ; \quad (\sigma_{\rho^{\text{q}}})^2 = \frac{T}{\Delta V}\left\langle \frac{1}{F^{\text{q}}(k)} \right\rangle_{\mathbf{k}} , \tag{5}$$
where T is the temperature and ΔV a volume cell in configuration space where $(\sigma_{\rho^{\text{q}}})^2$ is calculated.
An application of such scheme to nuclear matter is well suited to investigate the equilibrium variance of spacial isovector density correlations $(\sigma_{\rho^{\text{v}}})^2$ as a function of the symmetry energy. From an analogous application to open system, the isotopic distributions of clusters emerging from density ripples can be analysed.
- Second, in unstable conditions, both initial fluctuation seeds and intermittent fluctuation seeds injected at later times can yield an exponential growth of the intensity of response $(\sigma^{\text{q}}_k)^2$ over a growth time τ_k [17, 18].
An application of such scheme to nuclear matter is well suited to sample zero-sound propagation and investigate the instability growth rates $\Gamma_k = 1/\tau_k$. Analogous conditions in open system are suited to investigate the arising of clusterisation.

In the following, these steps are attended to.

2.1. Relating isovector fluctuations to the symmetry energy

To undertake a simulation on the propagation of isovector fluctuations, it is convenient to isolate them from the overwhelming effect of isoscalar fluctuations; those latter would in fact completely dominate the dynamics with their larger amplitude. Scalar terms are therefore suppressed in the potential so that stable conditions are imposed at any density ρ^0.

In this case, the fluctuation–dissipation relation Eq. (5) takes a form where the isovector-density variance is related to an effective symmetry energy, proportional to the symmetry energy E_{sym} at zero temperature [19]

$$(\sigma_{\rho^{\mathrm{v}}})^2 = \frac{T}{\Delta V}\left\langle \frac{1}{F^{\mathrm{v}}(k)} \right\rangle_{\mathbf{k}} \longrightarrow F^{\mathrm{v}}_{\mathrm{eff}} = \frac{T}{2\Delta V}\frac{\rho^0}{(\sigma_{\rho^{\mathrm{v}}})^2} = \frac{T}{2\Delta V}\frac{\rho^0}{\langle[\delta\rho_n(\mathbf{r}) - \delta\rho_p(\mathbf{r})]^2\rangle} \propto E_{\mathrm{sym}}(\rho^0)\ . \tag{6}$$

This relation can be solved numerically by calculating the isovector variance $(\sigma_{\rho^{\mathrm{v}}})^2$ in cells of edge size l, where l should be chosen large enough to minimise surface effects and better sample the volume symmetry energy (this value is found around $l = 2$fm).

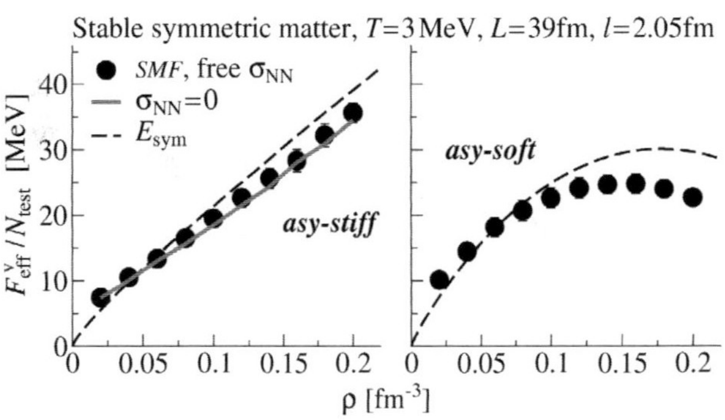

Figure 1. SMF calculation of the effective symmetry energy, scaled by N_{test}, compared to the analytic expression of the symmetry energy for different system densities. The collision term is either activated (and a free σ_{NN} is used) or suppressed ($\sigma_{\mathrm{NN}} = 0$). Two parametrisations of the symmetry energy, either asy-stiff (left panel), or asy-soft (right panel) are used.

In a first calculation, shown in Fig. 1, where the simplified SMF approach to solve the BL equation is used, even though the evolution of the effective symmetry energy $F^{\mathrm{v}}_{\mathrm{eff}}$ as a function of density reproduces correctly the shape of the analytic expression for the symmetry energy E_{sym}, the former and the latter differ of a factor equal to the number of test particles per nucleon N_{test} employed in the numerical sampling of the mean field:

$$F^{\mathrm{v}}_{\mathrm{eff}} \approx N_{\mathrm{test}} E_{\mathrm{sym}}\ . \tag{7}$$

The same result is obtained with or without the contribution of the residual term I_{UU}, indicating that explicit isovector terms are missing in the isovector channel in building fluctuations. This is not surprising because the residual term is not defined to build nucleon-nucleon correlation and the corresponding fluctuations. Those latter, even though reflecting the potential employed in the calculation, develop from initial fluctuation seeds which are related to the numerical noise in the sampling of the mean-field, which, on its turn, is related to N_{test}.

A second calculation, shown in Fig. 2, employs the BLOB approach, where the residual term is designed to explicitly introduce phase-space fluctuations from nucleon-nucleon correlations. In this case, the isovector variance is enhanced, but equilibrated nuclear matter is not a favourable condition to completely recover the large isovector variance of the analytic expectation. In particular, at small density ρ^0 collisional correlations are ineffective, at larger density collisions become rare and can hardly revive fluctuations and, in addition, the same noise issue already

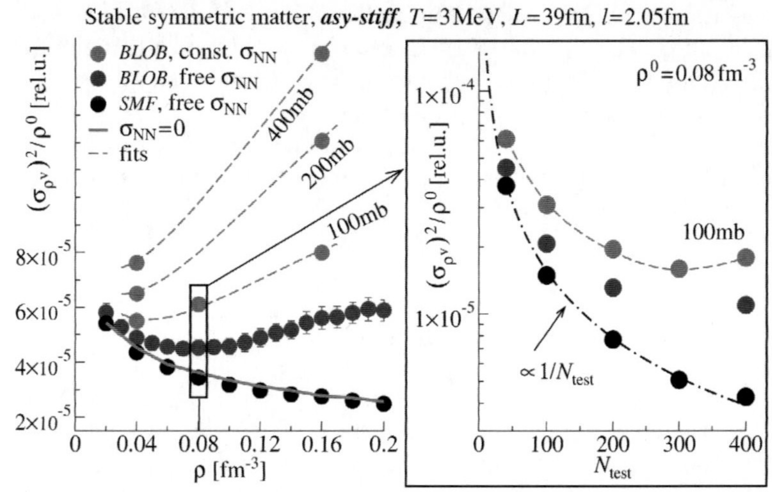

Figure 2. Isovector variance at equilibrium as a function of the system density for an asy-stiff parametrisation of E_{sym}. Calculated with BLOB, with different values of σ_{NN}, and compared with SMF, with the collision term either activated (with a free σ_{NN}) or suppressed. For the system density $0.08\mathrm{fm}^{-3}$, the inset shows the effect of varying N_{test} (see text).

observed in the SMF calculation is acting as a smearing contribution [17] with a dependence on N_{test}.

For test purposes, it is instructive to appreciate the effect of increasing the nucleon-nucleon cross section and, as a consequence, intensifying the nucleon-nucleon collision rate. This action enhances the effect of the BLOB fluctuating term in reviving nucleon-nucleon collisions, so as to prevail over the smearing effect of the mean-field noise.

A more consistent (but numerically costly) approach is reducing the numerical noise by refining the mean-field paving: this action, consisting in simply increasing N_{test}, also improves the results with respect to the SMF approach.

From these results and trends, we can expect that the limits imposed by the conditions of equilibrated nuclear matter could be actually overcame in out-of-equilibrium conditions, where nucleon-nucleon collision rates are definitely higher: this situation corresponds to the early stages of heavy-ion collisions, which are investigated in Sec. 3.1.

2.2. Relating zero-sound propagation to instability growth

To undertake a simulation of zero-sound propagation of collective modes in nuclear matter, the system should be prepared as initially homogeneous and at low temperature. If the temperature increases, a transition from zero to first sound may in fact occur [20, 21]. If unstable conditions are investigated, an amplification of the fluctuation amplitude is expected to trigger a catastrophic process. Early times are therefore better suited to compare to analytic expectations. If the zero-sound propagation is correctly sampled, isoscalar fluctuations should develop spontaneously with the correct amplitude, inducing the arising of a mottling pattern.

The analytic expectation can be defined in a linear-response approximation, assuming small deviations from a mean trajectory f^0; in this case, from the linearised Vlasov equation (where residual terms are suppressed) expressed as a function of the disturbance wave number k and frequencies ω_k, the corresponding dispersion relation [22, 23] can be extracted by applying self-consistency, as follows:

$$\omega_k f_k + \mathbf{k}\cdot\frac{\mathbf{p}}{m}f_k - \frac{\partial f^0}{\partial\epsilon}\frac{\partial U_k}{\partial\rho}\mathbf{k}\cdot\frac{\mathbf{p}}{m}\rho_k = 0 \;\longrightarrow\; 1 = \frac{g}{h^3}\frac{\partial U_k}{\partial\rho}\int\frac{\partial f^0}{\partial\epsilon}\frac{\mathbf{k}\cdot\mathbf{p}/m}{\omega_k+\mathbf{k}\cdot\mathbf{p}/m}\mathrm{d}\mathbf{p}\,. \quad (8)$$

At zero temperature, the eigenmodes f_k in the linearised Vlasov equation (left side of the arrow) depend on states near the Fermi level ϵ_{F} and the integral in the dispersion relation (right side

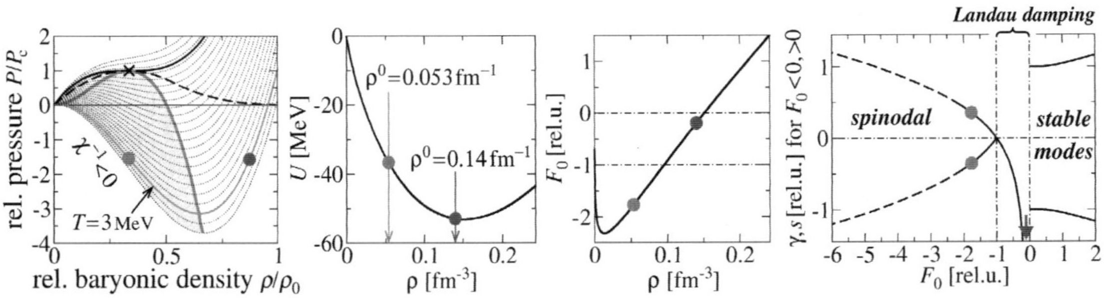

Figure 3. Two conditions, corresponding to the spinodal instability ($\rho^0 = 0.053\text{fm}^{-3}$) and Landau damping ($\rho^0 = 0.14\text{fm}^{-3}$) are indicated with dots in a sequence of four plots. From left to right: the equation-of-state landscape; the isoscalar mean-field potential as a function of the density; the Landau parameter F_0 as a function of the density; real (for $F_0 > 0$) and imaginary (for $F_0 < 0$) roots of the dispersion relation as a function of the Landau parameter F_0.

of the arrow) is restricted to the Fermi surface so that, introducing the Landau parameter $F_0(\mathbf{k}) = (3/2)(\rho^0/\epsilon_F)\partial_\rho U_k$, the dispersion relation can be written in terms of sound velocity $s = \omega_k/(kv_F)$ [18, 24]

$$1 + \frac{1}{F_0} = \frac{s}{2}\ln\left(\frac{s+1}{s-1}\right) . \tag{9}$$

As shown in Fig. 3, instabilities in zero-sound conditions can be tracked by placing the system in a location of the equation-of-state landscape where the incompressibility $\chi^{-1} \equiv \rho\frac{\partial P}{\partial \rho}$ is negative. In these conditions, corresponding to the spinodal instability and to $F_0(k{=}0) < -1$, the dispersion relation yields imaginary roots [25] which, by replacing $s \to i\gamma$, can be put in the

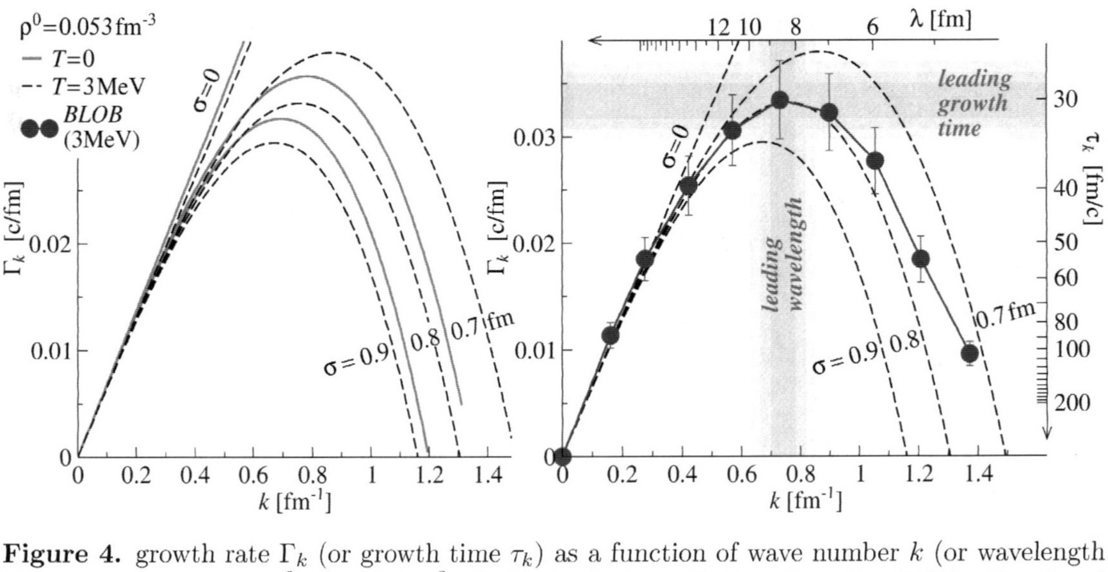

Figure 4. growth rate Γ_k (or growth time τ_k) as a function of wave number k (or wavelength λ) in nuclear matter at $\rho^0 = 0.053\text{fm}^{-3}$. (Left) Analytic predictions from Eq. (10) for $T = 0$ and $T = 3\text{MeV}$ and for different values of the Gaussian smearing factor σ. (Right) BLOB calculation from Eq. (11) for $T = 3\text{MeV}$ compared to the analytic predictions (the interaction, considering surface contributions, uses a σ in the region of 8 to 9fm): the resulting leading wavelength and growth time are indicated. Uncertainties are evaluated from the linear-response fit.

form

$$1+\frac{1}{F_0(k)}=\gamma\,\mathrm{arctan}\frac{1}{\gamma}\quad\longrightarrow\quad|\gamma|=\frac{|\omega_k|}{kv_\mathrm{F}}=\frac{1}{\tau_k k v_\mathrm{F}}\,. \tag{10}$$

Eq. (10) shows that disturbances of wave number k get amplified with a growth time τ_k and a corresponding growth rate $\Gamma_k = 1/\tau_k$.

As shown in the analytic calculations of Fig. 4 (left panel) for zero temperature, the response intensity at zero sound should present the following evolution of the growth rate Γ_k as a function of the k number, or with the corresponding wavelength λ.

- For small k: Γ_k tends to increase (decrease) linearly with k (λ) because the more matter has to be relocated, the longer time it takes.
- For large k: small wavelengths λ are excluded as a function of the interaction range. This ultraviolet cutoff can be described as a Gaussian smearing [27, 26] σ of the mean-field potential in configuration space $U \otimes g(k)$ with $g(k) = \mathrm{e}^{-\frac{1}{2}(k\sigma)^2}$ which reduces Γ_k with k.
- The combination of these opposite behaviours produces a maximum which corresponds to the fastest growing disturbance, i.e. the leading k mode and leading wavelength.

To explore the equation of state (especially in applications to heavy-ion collisions), it is convenient to have a finite value of T. Still, as mentioned, T should be low in order to avoid transitions to first-sound. It is possible to introduce a finite temperature T rather than working at $T = 0$ through a low-temperature expansion of the chemical potential ν. A reduction of Γ_k with T is the result of this action.

These modifications to take into account the ultraviolet cutoff and to introduce a finite temperature impose to replace $F_0(k)$ by an effective Landau parameter [27] $\widetilde{F}_0(k,T) = (\mu(T)/\epsilon_\mathrm{F})F_0 g(k)$ in eq. (10) in order to obtain the analytic response in Fig. 4 for a finite temperature.

The corresponding numerical calculations employ the BLOB approach to extract the response intensity $\sigma_k^2(t)$, i.e. the amplitude of the isoscalar fluctuation of a mode k from the Fourier transform of the space density (see ref. [2] for details). Fig. 5 studies the evolution of the ratio $\tilde{\sigma}_k^2(t) = \sigma_k^2(t)/\sigma_k^2(t=0)$.

From the analysis of the response intensity we infer some general results which can be transposed from nuclear matter to heavy-ion collisions.

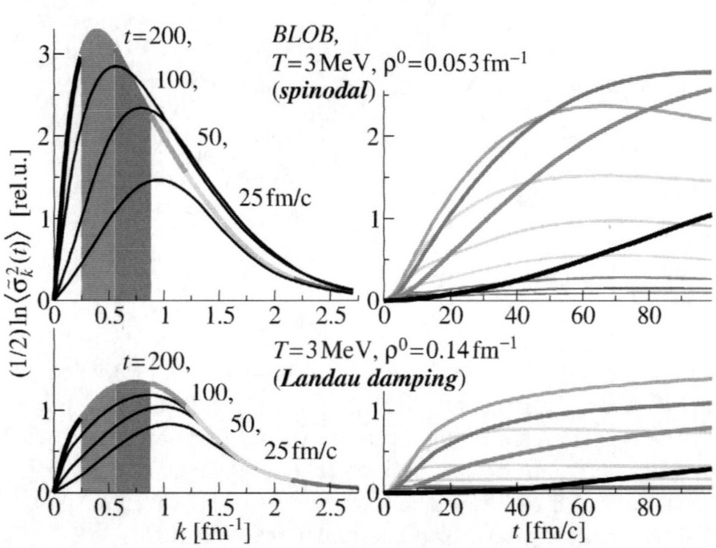

Figure 5. Average response intensity as a function of k (left column) and time t (right column), calculated with BLOB in conditions of spinodal-instability (top row) and Landau damping (bottom row). The different curves in the left column correspond to different time instants, while the different curves in the right column correspond to different k intervals (their colours corresponds to the coloured segments in the left panel).

IOP Conf. Series: Journal of Physics: Conf. Series **1014** (2018) 012008 doi:10.1088/1742-6596/1014/1/012008

- In unstable (spinodal) conditions, calculated at $\rho^0 = 0.053$ fm^{-3} fluctuations get rapidly amplified as expected.
- Even outside of the spinodal instability, in calculations at $\rho^0 = 0.14$ fm^{-3}, which correspond to Landau-damping conditions, the response intensity reaches significant amplitudes.
- The response intensity saturates rather early due to the combination of large k into small k. In heavy-ion collisions, a similar effect manifests in the recombination of small emerging clusters into larger fragments and in an overall mean-field resilience effect (investigated in detail in ref. [28]).

To the analytic expectation for the growth-rate as a function of k, Fig. 4 (right panel) overlaps the corresponding numerical calculation obtained with BLOB for the same conditions and interaction properties, when taking the average over several stochastic dynamical paths,

$$\Gamma_k = \frac{1}{2}\frac{\partial}{\partial t}\ln \prec \tilde{\sigma}_k^2(t) \succ \ . \tag{11}$$

The achieving of a close correspondence between the BLOB calculation and the analytic expectation, as shown in Fig. 4 ensures that the model succeeds in describing the fluctuation phenomenology consistently and, in particular, fluctuations develop with the correct amplitude. After showing that this correspondence is achieved, some quantitative results can be extracted from the calculation. The leading wavelength ranges between 8 to 9 fm. By analogy to nuclear matter, we can expect for heavy-ion collisions that fragments and clusters should arise in the region of Neon. The corresponding separation time in heavy-ion collisions should not only take into account the growth time of the leading k modes, calculated in Fig. 4, but also the time needed in the collision to generate a low-density phase where instabilities may develop, and the overall effect of the kinematics in an open system.

3. Dynamics of clusterisation in open systems, applications to heavy-ion collisions

The composition of the two schemes presented at the beginning of Sec. 2 and applied to open systems provides an analysis of the isovector and isoscalar properties of clustering in stable and unstable condition, in connection with analytic expectations from Fermi fluids.

3.1. Spinodal clusters from density ripples at Fermi energies

An analysis which directly corresponds to the above calculations in nuclear matter is dedicated to the properties of emerging clusters in an open system at low density, produced in a heavy-ion collision. Even before that clusters separate into well identified blobs of matter in configuration space, we can track the evolution of potential ripples developing in the bulk since early times. In analogy to independent fragments, we can attribute to those inhomogeneities a mean radius and a density averaged over the corresponding potential well in order to obtain mass and element numbers A', Z'. An indicative expectation for isotopic yield distributions may be obtained by analogy to eq. (5) as

$$Y \approx \exp[-(\delta^2/A')\, C_{\rm sym}(\rho)/T] \ . \tag{12}$$

These yields of forming fragments reflect the isovector fluctuations. As already argued in Sec. 2.1, while in equilibrated nuclear matter fluctuations are hard to revive and entertain due to low collision rates, in open systems fluctuations are initially built out of equilibrium, so that collision rates are larger. As a consequence, the isovector fluctuations variance results more consistent with analytic expectations, as illustrated in Fig. 6 (in the frame) for the system ^{136}Xe+^{124}Sn at 32 AMeV (central collisions), where the isotopic distribution of forming fragments is calculated with BLOB and compared to the expectation of eq. (12).

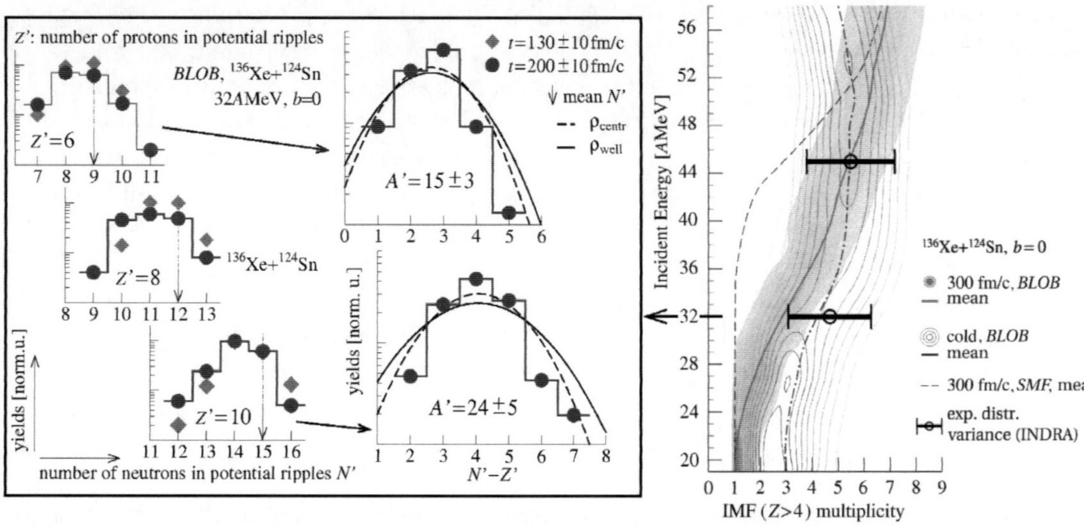

Figure 6. In the frame, BLOB calculation of the isotopic production in central collisions in ^{136}Xe+^{124}Sn at 32 *A*MeV. (Left column in the frame) mass distributions of forming isotopes of C, O and Ne at early times $t = 130$ and 200 fm/c, before separation into independent fragments. (Right column in the frame) Distributions of isotopic variances in potential ripples containing N' neutrons and Z' protons for the most probable configurations for forming clusters in the regions around C ($Z' = 6$ and $A' = 15$) and Ne ($Z' = 10$ and $A' = 24$) at $t = 200$ fm/c. For comparison, analytic distributions are plotted, as a function of the density measured in the minimum of the potential ripples $\rho_{\rm centr}$ or averaged over the potential ripples $\rho_{\rm well}$. The right panel shows the evolution of the multiplicity of fragments produced in central collisions in ^{136}Xe+^{124}Sn at various incident energies, calculated with BLOB (colour map) at $t = 300$ fm/c when fragments separate, and at the end of the whole decay sequence, calculated by BLOB+Simon (grey contours). The mean values and distribution widths are compared with experimental data from INDRA [31, 30] (the bar indicates the variance of the multiplicity distribution).

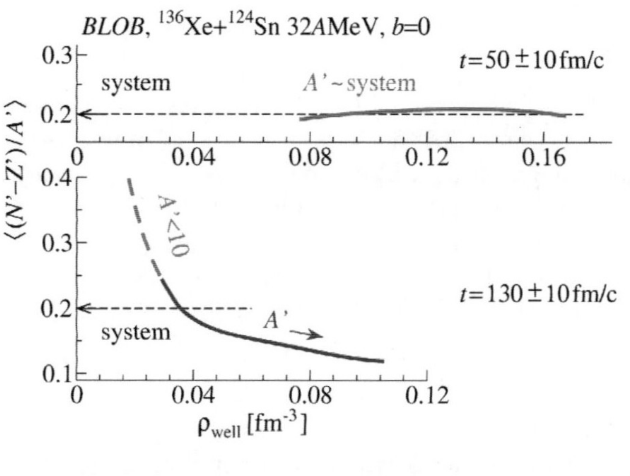

Figure 7. The isotopic properties of fragments produced in central collisions in ^{136}Xe+^{124}Sn at 32 *A*MeV is shown to evolve according to a process of isospin distillation. The average isospin content in potential ripples as a function of $\rho_{\rm well}$ is calculated with BLOB before the onset of the fragmentation process at $t = 50$fm/c, or when the process is almost achieved at $t = 130$fm/c.

Fig. 7 extends the study of the isospin content of forming clusters and fragments to the whole range of sizes produced in the process, and illustrates the corresponding average isospin content at a very early time ($t = 50$fm/c), when the system is still rather homogeneous, and at a later time ($t = 130$fm/c), when inhomogeneities have been built, as a function of the local density $\rho_{\rm well}$ associated to the density ripple. The larger neutron enrichment of the more volatile phase

IOP Conf. Series: Journal of Physics: Conf. Series **1014** (2018) 012008 doi:10.1088/1742-6596/1014/1/012008

signs the onset of a distillation process [24].

The right panel of Fig. 6 extends the survey of fragment production to a large incident-energy range, examining the multiplicity of fragments with $Z > 4$. In this case, fragments are studied at $t = 300\text{fm/c}$, when they are completely separated but still highly excited, and at the end of the full decay sequence (where the Simon transition-state model [29] was employed to complete the decay with sequential evaporation). The BLOB+Simon calculation yields mean values and widths which are in agreement with experimental data from INDRA [31, 30]. The transition from fusion to multifragmentation occurs below $30A$MeV and the dominant spinodal mechanism is gradually replaced by vaporisation beyond $45A$MeV [11]. On the other hand, the simplified BL approach to introduce fluctuations used in SMF results less efficient in building fluctuations, and does not succeed in describing the transition from fusion to multifragmentation at the correct incident energy.

3.2. Light systems

Heavy-ion collisions and nuclear matter involve anditional clustering processes, where alpha and light charge particles are formed. Such mechanism, different from dynamical instabilities, would require the explicit inclusion of further correlations in the present formalism. Light charged particles related to nuclear clustering have in fact too small size, exceeding the ultraviolet cutoff of the dispersion relation, so that they can not belong to the unstable multipole modes which characterise spinodal fragmentation. Solutions for an explicit treatment of cluster formation are proposed in refs. [32, 5].

The ability of the stochastic approach of BLOB in handling fragment formation and introducing correlations bejond the level of kinetic equations can be considered a first approximation to light clusters, which are essentially emerging from dynamical fluctuations. Blob already takes charge of a fraction of the light cluster production and a sequential-decay afterburner (Simon [29] was used) adds the missing fraction from the decay of heavier elements. A comparison with data from ref. [33] is presented in Fig. 8 for the system ^{12}C+^{12}C at 62 AMeV.

At this level of comparison, the production rates of the light masses in the isobaric distribution seems consistently described. Such production is fed by the most central impact parameters and fades with periferal configurations, which are related to the heavier side of the mass distribution.

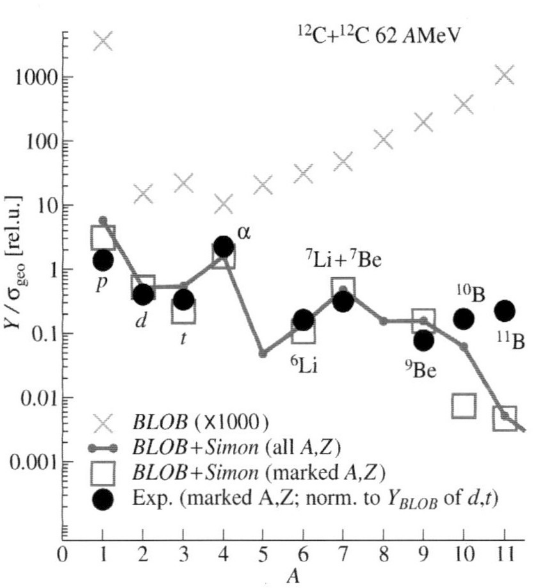

Figure 8. Clusters and fragment yields in ^{12}C+^{12}C at 62 AMeV, divided by the geometric cross section, are shown as calculated with BLOB till the last registered split in a time window of 80 to 140 fm/c (BLOB, scaled by 1000) and after the full decay sequence, for which the model Simon [29] was employed (BLOB+Simon). A free $\sigma_{\rm NN}$ and an asy-stiff form for the symmetry energy are used. Experimental yields from ref. [33] for some isotopes named on the plot, are indicated with dots; data are normalised so that the sum of d and t yields equals the corresponding calculated quantity. Squares are calculations corresponding to the isotopes that have been measured.

3.3. Fluctuations and transparency at intermediate energy

The connection between fragment production and hydrodynamic properties like the characterisation of the flow [34] at intermediate energies is strictly related to the treatment of fluctuations.

At intermediate energy the inclusion of fluctuations has two antagonist effects: on the one hand, it enhances the fragmentation of the system, on the other hand it reduces the directed flow. This effect can be studied in the comparisons of Fig. 9, for the collision ^{197}Au+^{197}Au at 100 AMeV for an impact parameter of 7fm. The simulation is performed with three approaches, SMF without and with a collision term (constant $\sigma_{\rm NN} = 40$mb) and BLOB (also with $\sigma_{\rm NN} = 40$mb), using identical parameters for the mean field as defined in ref. [35]. The SMF approach describes the outward deflection of the trajectory imparted by the directed flow, which is absent in the SMF description without collision term. The BLOB approach exhibits a reduced directed flow with respect to SMF, because it competes with the production of fragments and clusters. This latter, due to the Langevin fluctuations, results in a large variety of very different fragment configurations; two of those are shown, one where the fragmentation of the quasi-target and the quasi-projectile is observed (bottom row in the density map, left), the other where the emitting source is situated at midrapidity (bottom row in the density map, right).

A quantitative study of the flow is illustrated in the right panel of Fig.9. For a simulation where the collision rate is identical (due to using the same constant $\sigma_{\rm NN}$), the larger fragmentation rate is reflected in a smaller slope for the directed flow as a function of reduced rapidity. This is a rather general example to spots the main difference between a simplified description of fluctuations and a BL approach solved in full phase space, when applied to intermediate energies.

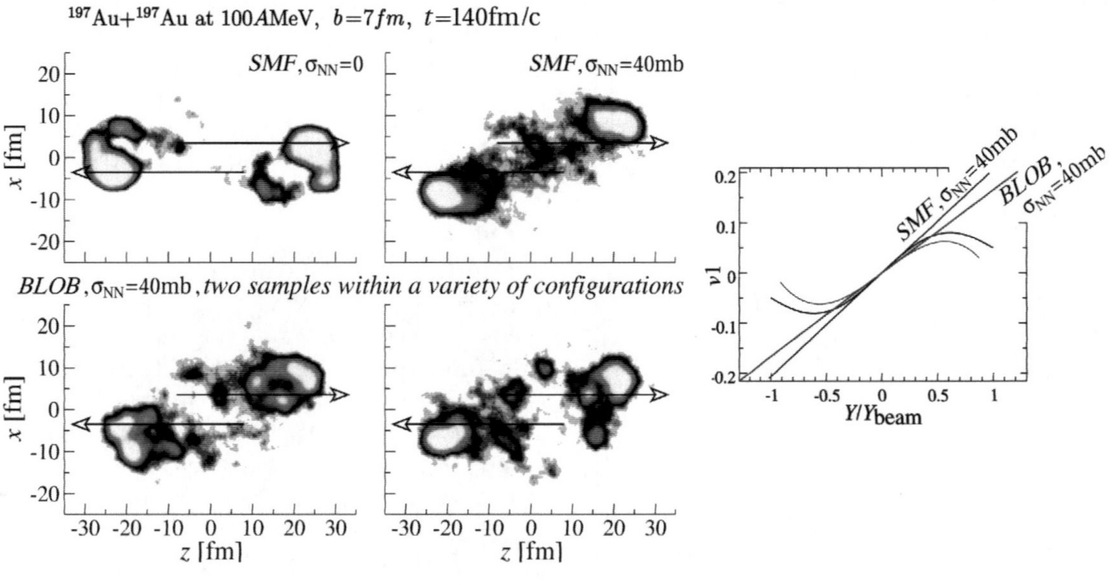

Figure 9. Simulation of the collision ^{197}Au+^{197}Au at 100 AMeV for an impact parameter of 7fm, studied at the time t = 140 fm/c and simulated through three approaches: SMF without collision contribution (top, left in the density map), SMF (top, right in the density map) and BLOB (bottom row in the density map). The arrows indicate the direction of the target and projectile; their origins indicate the centres of target and projectile at the initial time t = 0 for the simulation. (Right panel) Directed flow as a function of reduced rapidity for SMF and BLOB. The corresponding slope at zero reduced rapidity is indicated.

4. Conclusions

The process leading a fermionic system, like nuclear matter or violent nuclear reactions, to separate into clusters and fragments is reviewed in some significant steps.

Among more possible strategies to address the problem (see introduction), we have chosen to extend a mean-field description to include fluctuations in a Boltzmann-Langevin framework [2] and suggested some scheme for testing the transport approach on analytic expectations.

The approach is particularly adapted to describe mean-field inhomogeneities, like the spinodal decomposition, but also conditions like Landau damping and other collective behaviours. It reduces to an approximation when the formation of the light clusters should be described because no explicit cluster contributions are so far treated. For larger clusters and intermediate-mass fragments in general, the approach yields quantitative predictions for the production of fragments in a broad range of situations and mechanisms encountered at Fermi energies [11, 36, 37]. To complete the survey, a further application to relativistic spallation has been undertaken elsewhere [28].

References

[1] Pines D and Nozieres P 1966 *The Theory of Quantum Liquids* (New York: Benjamin)
[2] Napolitani P and Colonna M 2017 *Phys. Rev.* C **96** 054609
[3] Ono Akira and Randrup J 2006 *Eur. Phys. J.* **A30** 109
[4] Ono Akira, Horiuchi H, Maruyama T and Ohnishi A 1992 *Phys. Rev. Lett.* **68** 2898;
[5] Ono Akira 2016 *EPJ Web Conf.* **122** 11001
[6] Feldmeier H, 1990 *Nucl. Phys.* **A515** 147; Feldmeier H, Bieler K and Schnack J 1995 *Nucl. Phys.* **A586** 493
[7] Balescu R 1976 *Equilibrium and Non-equilibrium Statistical Mechanics* (New York: Wiley)
[8] Ayik S and Grégoire C 1988 *Phys. Lett. B* **212** 269
[9] Lacombe L, Suraud E, Reinhard P -G and Dinh P M 2016 *Ann. Phys.(NY)* **373** 216
[10] Reinhard P -G and Suraud E 1992 *Ann. of Physics* **216** 98
[11] Napolitani P and Colonna M 2013 *Phys. Lett. B* **726** 382
[12] Napolitani P and Colonna M 2012 *EPJ Web of Conf.* **31** 00027
[13] Rizzo J, Chomaz Ph and Colonna M 2008 *Nucl. Phys.* **A806** 40
[14] Colonna M et al. 1998 *Nucl. Phys.* **A642** 449
[15] Guarnera A, Colonna M and Chomaz Ph. 1996 *Phys. Lett. B* **373** 267
[16] Baran V, Colonna M, Greco V and Di Toro M 2005 *Phys. Rep.* **410** 335
[17] Colonna M, Burgio G F, Chomaz Ph, Di Toro M and Randrup J 1993 *Phys. Rev.* C **47** 1395
[18] Colonna M, Chomaz Ph and Randrup J 1994 *Nucl. Phys.* **A567** 637
[19] Colonna M 2013 *Phys. Rev. Lett.* **110** 042701
[20] Larionov A B, Cabibbo M, Baran V and Di Toro M 2000 *Phys. Rev.* C **61** 064614
[21] Kolomietz V M, Plujko V A and Shlomo S 1996 *Phys. Rev.* C **54** 3014
[22] Landau L D 1957 *Sov. phys. JETP* **5** 101
[23] Kalatnikov I M and Abrikosov A A 1958 *Sov. phys. JETP* **6** 84
[24] Chomaz Ph, Colonna M and Randrup J 2004 *Phys. Rep.* **389** 263
[25] Pomaranchuk I Ia 1959 *Sov. phys. JETP* **8** 361
[26] Kolomietz V M and Shlomo S 1999 *Phys. Rev.* C **60** 044612
[27] Chomaz Ph and Colonna M 1994 *Phys. Rev* C **49** 1908
[28] Napolitani P and Colonna M 2015 *Phys. Rev.* C **92** 034607; 2016 *Il Nuovo Cimento* **39 C** 384
[29] Durand D 1992 *Nucl. Phys.* **A541** 266
[30] Ademard G et al. (INDRA Coll.) 2014 *Eur. Phys. J. A* **50** 33
[31] Gagnon-Moisan F et al. (INDRA coll.) 2012 *Phys. Rev C* **86** 044617
[32] Danielewicz P and Bertsch G F 1991 *Nucl. Phys.* **A533** 712
[33] De Napoli M 2012 *Phys. Med. Biol.* **57** 7651
[34] Reisdorf W and Ritter H G 1997 *Annu. Rev. Nucl. Part. Sci.* **47** 663
[35] Xu Jun et al. 2016 *Phys. Rev.* C **93** 044609
[36] Colonna M, Napolitani P and Baran V 2017 *Mean-Field Instabilities and Cluster Formation in Nuclear Reactions* in *Nuclear Particle Correlations and Cluster Physics* (World Scientific), pp. 403424
[37] Napolitani P and Colonna M 2016 *EPJ Web of Conf.* **117** 07008

Mass and angular distributions of the reaction products in heavy ion collisions.

A K Nasirov[1,2], G Giardina[3], G Mandaglio[4,5,6], B.M. Kayumov[2], R.B. Tashkhodjaev[2,7]

[1] JINR - Bogoliubov Laboratory of Theoretical Physics, Dubna, Russia
[2] Institute of Nuclear Physics, UzAS, Tashkent, Uzbekistan
[3] Dipartimento di Scienze Matematiche e Informatiche, Scienze Fisiche e Scienze della Terra, University of Messina, Messina, Italy
[4] Dipartimento di Scienze Chimiche, Biologiche, Farmaceutiche ed Ambientali, University of Messina, Messina, Italy
[5] INFN Sezione di Catania, Catania, Italy
[6] INFN Laboratori Nazionali di Frascati, Frascati, Italy
[7] Inha University in Tashkent, Tashkent, Uzbekistan

Abstract. The optimal reactions and beam energies leading to synthesize superheavy elements is searched by studying mass and angular distributions of fission-like products in heavy-ion collisions since the evaporation residue cross section consists an ignorable small part of the fusion cross section. The intensity of the yield of fission-like products allows us to estimate the probability of the complete fusion of the interacting nuclei. The overlap of the mass and angular distributions of the fusion-fission and quasifission products causes difficulty at estimation of the correct value of the probability of the compound nucleus formation. A study of the mass and angular distributions of the reaction products is suitable key to understand the interaction mechanism of heavy ion collisions.

1. Introduction

The mixing the fusion-fission and quasifission contributions due to overlap their mass-angle distributions in measured data leads to ambiguities at estimation of the fusion probability. The significance of this mixing depends on the total mass of colliding nuclei and their mass-asymmetry. If mass and charge numbers of the light nucleus is much smaller than ones of heavy nucleus ($A_1 << A_2$) the colliding system is very mass asymmetric. In this case overlap of the mass distributions of the fusion-fission and quasifission is very small since during evolution of dinuclear system, which is formed after capture of projectile by target-nucleus, the nucleon transfer from light fragment to the heavy one is hindered by the barrier at the Businaro-Gallone point of the driving potential. But the mass asymmetry reactions used to synthesize superheavy elements is between the Businaro-Gallone point and mass symmetric region ($A_1 \approx A_2$). In this area, the gradient of the potential energy surface produces the forces causing diffusion of nucleons from heavy fragment to light fragment.

As a result, the part of mass distribution of quasifission fragments in the mass asymmetric region increases. Therefore, it is actual to study the mass distribution of the fission-like products to find ways to separate quasifission from the fusion-fission products. The sketch of the sequences of nuclear reaction mechanisms in heavy ion collisions is shown in the left part of Fig. 1. Its right

IOP Conf. Series: Journal of Physics: Conf. Series **1014** (2018) 012009 doi:10.1088/1742-6596/1014/1/012009

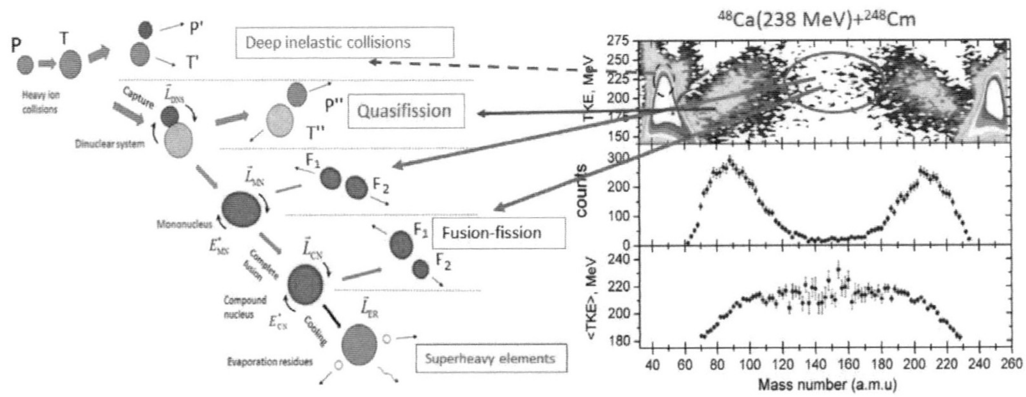

Figure 1. (Color on-line) The sketch of the sequences of nuclear reaction mechanisms in heavy ion collisions (left part) and their relations between experimental data of the mass and energy distribution of binary fragments (right part) in case of the ^{48}Ca+^{248}Cm reaction.

part is the mass and energy distributions measured for the deep-inelastic collisions, quasifission, and fusion-fission reactions in the ^{48}Ca+^{248}Cm reaction []. The aim of this work is to establish the mechanism preceding to the formation of the group of products of reaction. It is seen that there is overlap between mass distributions of deep-inelastic collision and capture event products which are projectile- and target-like. They can be separated due to their different kinetic energies (see right part of Fig. 1). At the capture events the full momentum transfer takes place and their products have smaller kinetic energies than the deep-inelastic collision products having not completely damped momenta. The difference between these two type reactions is explained in Fig. 2. In this figure the full momentum transfer corresponds to capture the curve of the path showing the decrease of the kinetic energy into potential well.

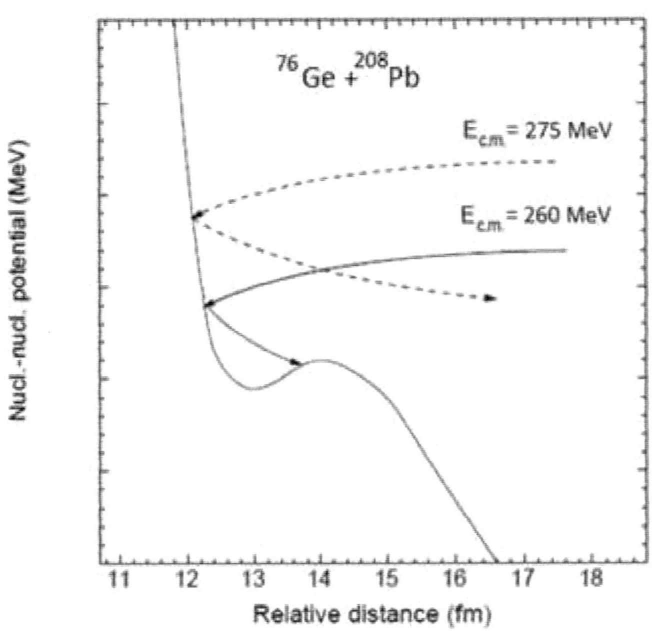

Figure 2. The nucleus-nucleus potential and damped kinetic energy in the entrance channel in the ^{48}Ca+^{248}Cm reaction for the head-on collision ($L = 0$).

IOP Conf. Series: Journal of Physics: Conf. Series **1014** (2018) 012009 doi:10.1088/1742-6596/1014/1/012009

The paths of the kinetic energy collected in the potential well are capture events which lead to formation of the dinuclear system. Its evolution is related by the nucleon exchange between fragments and decay through quasifission barrier which is the border of the potential well. The arrows "c" and "d" in Fig.3 show some quasifission events with different charge distributions. The arrow "b" shows the flow of events directed to complete fusion when all nucleons from the light fragment transferred to heavy one.

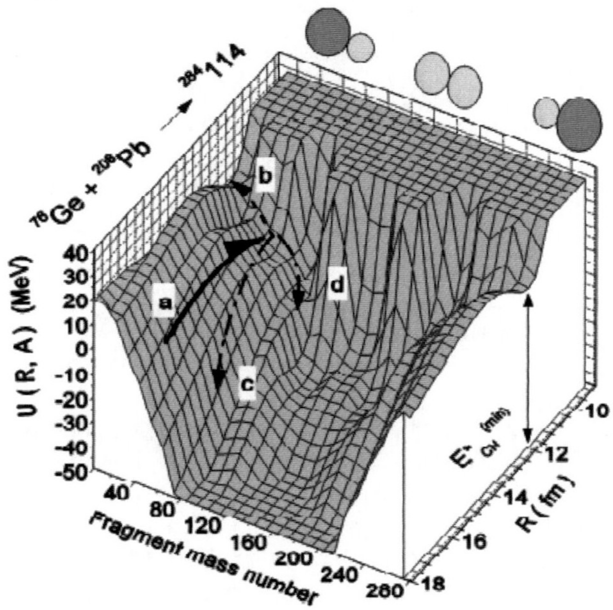

Figure 3. The nucleus-nucleus potential and damped kinetic energy in the entrance channel in the $^{48}Ca+^{248}Cm$ reaction for the head-on collision ($L = 0$).

So, the capture events presented by arrow "a" in Fig. 3 are distributed to the complete fusion "b" and quasifission "c" and "d" events. The evolution of the DNS demonstrates competition between complete fusion and quasifission. The transfer of all nucleon from the light fragment to heavy nucleus leads to formation of the rotating mononucleus which is excited too. Rotation of the non-equilibrated mononucleus causes decreasing the fission barrier. Therefore, at definite value of the orbital angular momentum the fission barrier disappears completely. In this case, the non-equilibrated mononucleus becomes unstable and undergoes to fission from some excited state. The splitting of mononucleus before reaching its equilibrium state is called fast fission which produces binary fragments (see the left part of Fig.1). The mass distribution of the fragments of the fast fission can be mixed with the ones of the fusion-fission and quasifission. The compound nucleus is equilibrated mononucleus having the non-zero fission barrier can registered as superheavy element if it must survive against fission emitting neutrons and charged particles (see left part of Fig.1). The study of correlations between mass and angular distribution of fragments of full momentum transfer reactions expected to be important in separation of the pure fission fragments of the compound nucleus with a compact shape from the quasifission products when they are mixed.

The quasifission is full-momentum transfer reaction. Mainly five collective motions related with the evolution of DNS formed at capture should be analyzed:

i) the nucleon exchange and mass drift (change of the mass asymmetry) between fragments takes place continuously with intensity depending on the shell structure of interacting nuclei and on the dissipated energy into nucleon motion from the bombarding energy;

ii) at the same time the DNS rotates with the angular velocity which depends on the initial orbital angular momentum and its moment of inertia;

iii) the relative distance between DNS fragments fluctuates that is allowed by the potential

IOP Conf. Series: Journal of Physics: Conf. Series **1014** (2018) 012009 doi:10.1088/1742-6596/1014/1/012009

well of the nucleus-nucleus interaction potential;

iv) deformation and surface vibration of the DNS fragments;

v) "butterfly" mode of vibration of the DNS fragments around axis perpendicular to the line connecting their centres-of-mass.

2. Mass and charge distribution evolution in the interacting system

The relation between the mass-angle distribution characteristics and landscape of the total potential energy as a function of the DNS charge and mass asymmetry show an important role of the entrance channel parameters and compound nucleus fissilities in the overlap of the characteristics of the reaction products formed by different mechanisms.

The dominating role of the quasifission process is noted by experimentalists too. The quasifission process is a hindering factor to complete fusion. For this reason, the cross section for producing a heavy evaporation residue (ER) is calculated by taking into account fusion probability $P_{\rm CN}$ and the expression can be written as

$$\sigma_{\rm ER} = \sigma_{\rm cap} P_{\rm CN} W_{\rm sur}, \quad (1)$$

where $\sigma_{\rm cap}$ is the capture cross section and $W_{\rm sur}$ is the probability that the compound nucleus (CN) survives against fission. Experimentally, the capture cross section is defined as the sum of the quasifission ($\sigma_{\rm QF}$), fusion-fission ($\sigma_{\rm CN-fiss}$), and ER ($\sigma_{\rm ER}$) cross sections,

$$\sigma_{\rm cap} = \sigma_{\rm QF} + \sigma_{\rm CN-fiss} + \sigma_{\rm ER}. \quad (2)$$

They assume that in the symmetric region of the fragment masses ($A/2 \pm 20$) fusion-fission process coexists with quasifission [1]. The experimental mass distribution of the fission-like products is fitted with the set of the Gaussian functions to separate the contributions of the fusion-fission products from the ones of quasifission. The mass yield of the fusion-fission process obtained as difference between experimental spectra and quasifission peak in the mass symmetric region [1,2]. $P_{\rm CN}$ is defined as the probability for compound nucleus formation from the configuration of two nuclei in contact. In the reactions with massive nuclei the ER cross section is less than a few nanobarns and contributes insignificantly to the fusion cross section. Thus we can estimate the fusion probability using the measured mass-energy distributions as the ratio between the number of events attributed to fusion-fission and all fissionlike fragments. Therefore, it is important to study in detail the reaction mechanisms producing the fissionlike fragments and to be able to separate the contributions from the different mechanisms to the registered yield of the reaction products. The ratio of the different contributions into yield of the fissionlike products depends on the mass asymmetry of colliding nuclei, orientation angles of their axial symmetry, beam energy and orbital angular momentum in the entrance channel.

It is obvious the differences between the values of $P_{\rm CN}$ extracted from the measured data of the capture and complete fusion events depend on how are correctly estimated capture and fusion cross sections from the measured data. Therefore, the reliable experimental determinations of $P_{\rm CN}$ and consequently the understanding the entrance channel effect on the $P_{\rm CN}$ values are strongly related to the choice of assumptions for the data analysis.

The experimental results of the mass-total kinetic energy matrices for the ^{36}S, ^{48}Ca+^{238}U reactions at energies close to the Coulomb barrier taken from [3] are presented in Fig. 4. The mass distributions for fission-like fragments presented by open circles are sum of the events which are inside the region limited by thick lines on mass-total kinetic energy matrices in the top of Fig. 4. The determination of the fusion probability

$$P_{\rm CN} = \frac{\sigma_{\rm fus}}{\sigma_{\rm cap}} = \frac{\sigma_{\rm CN-fiss} + \sigma_{\rm ER}}{\sigma_{\rm QF} + \sigma_{\rm CN-fiss} + \sigma_{\rm ER}}. \quad (3)$$

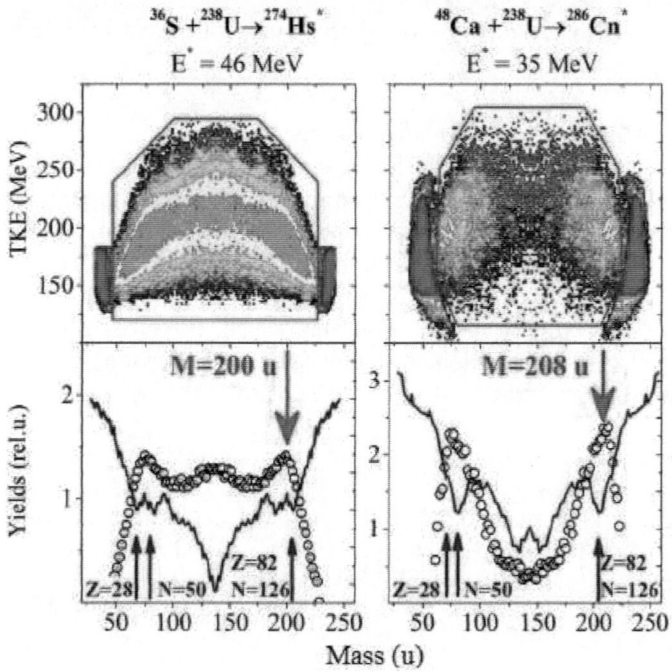

Figure 4. (Color on-line). The experimental mass-total kinetic energy matrices for the ^{36}S, ^{48}Ca+^{238}U reactions at energies close to the Coulomb barrier taken from [3]; bottom: open circles are mass distributions for fission-like fragments inside the region limited by thick lines on mass-TKE matrices; solid lines are the potential energies as a function of mass.

from the mass distributions for fission-like fragments depends on the assumption used at separation of the quasifission and fusion-fission events. This procedure gives the ratio $\sigma_{\rm CN-fiss}/\sigma_{\rm QF}$. We should note that the position of the border of the region in the top of Fig. 4 under discussion is incorrect since a lot of events of the quasifission related with the formation of the products with mass numbers $A_1 < 44$ and $A_2 > 230$ are ignored. So, the incorrect choice of the region borders leads to decrease the denominator of Eq. 3. As a result the experimental fusion probability is already overestimated. The verify this statement we can analyze the results of theoretical charge (mass) distribution for the quasifission and fusion-fission products calculated by the master equation [4, 5] for the nucleon transfer between interacting nuclei. It is seen the overlap of these two yields in the range of mass $80 < A_1 < 120$. The calculation has been made for the $E_{\rm lab} = 185$ MeV, orbital angular momentum value $L = 20\hbar$ and orientation angle $\alpha_2 = 45°$. The amount of the fusion-fission events is small in case of the ^{48}Ca+^{238}U reaction since the fusion probability .

Mass and charge distributions of the quasifission products $Y_{\rm qf}(Z, A)$ are calculated by solving the transport master equation [4,5] for the probability of the charge distribution P_Z in the DNS fragments:

$$\frac{\partial}{\partial t}P_Z(E^*_Z,t) = \Delta^{(-)}_{Z+1}(E^*_Z)P_{Z+1}(E^*_Z,t) + \Delta^{(+)}_{Z-1}(E^*_{Z-1})P_{Z-1}(E^*_Z,t) - (\Delta^{(-)}_Z(E^*_Z) + \Delta^{(+)}_Z(E^*_Z) + \Lambda^{qf}_Z(E^*_Z))P_Z(E^*_Z,t), \quad (4)$$

where $Z = 2, 3, ..., Z_{tot} - 2$ and the transition coefficients $\Delta^{(\pm)}_Z$ determine the probability of nucleon transfer between interacting nuclei "P" and "T" of the dinuclear system characterized with the charge numbers Z and $Z_{\rm tot} - Z$, $Z_{\rm tot}$ is the total charge number of the system; E^*_Z is its excitation energy which is determined by the initial energy of collision $E_{\rm c.m.}$, the minimum value of the potential well $V^{(Z)}_{\rm min}$ in the nucleus-nucleus interaction and the energy balance of the nucleon transfer as a difference of the DNS total energy U values of for the entrance channel

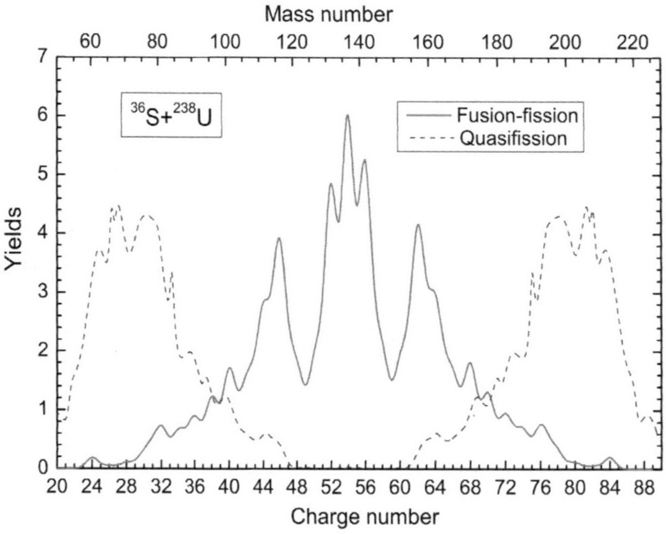

Figure 5. (Color on-line). The mass distributions for fusion-fission (solid line) and quasifission (dashed line) products calculated by solution of the transfer master equation [4, 5] for the ^{36}S+^{238}U reaction at beam energy $E_{\rm lab}$ = 185 MeV.

and the charge (mass) asymmetry $Z(A)$:

$$E^*(Z,A) = E_{\rm c.m.} - V_{\rm min}^{(Z_P,A_P)}(R_{\rm min}) + \Delta U(Z,A), \tag{5}$$

where

$$\begin{aligned}\Delta U(Z,A) &= U(Z_P,A_P) - U(Z,A) \\ &= B_1(Z_P,A_P) + B_2(Z_T,A_T) + V_{\rm min}^{(Z_P,A_P)} \\ &- (B_1(Z,A) + B_2(Z_{\rm CN}-Z, A_{\rm tot}-A) + V_{\rm min}^{(Z,A)}. \end{aligned}\tag{6}$$

So, the dependence the DNS excitation energy on the charge asymmetry is related with the change of its intrinsic energy by the change of mass and charge numbers of its constituents from the ones of projectile and target nuclei. The corresponding mass numbers A are found by the minimization of the total energy of dinuclear system as a function of A; $\Lambda_Z^{\rm qf}$ is the Kramer's rate for the decay probability of the dinuclear system into two fragments with charge numbers Z and $Z_{\rm tot} - Z$ and it is proportional to $\exp(-B_{\rm qf}(Z)/T)$ where $B_{\rm qf}(Z)$ is quasifission (pre-scission) barrier against decay and T is the effective temperature of the system. Details of calculation of $B_{\rm qf}$ and T are presented in Ref. [9].

The peculiarities of the mass distribution related with the appearance of the strong yield of the reaction products with the definite charge numbers are determined by the dependence of the transition coefficients $\Delta_Z^{(\pm)}$ on the single-particle structure of the proton and neutron subsystems in the interacting nuclei.

The transition coefficients $\Delta_Z^{(\pm)}$ depends on the energy (ε_i), spin (J_i) and occupation numbers (n_i of the single-particle states of the nuclei with charge number Z and $Z_{\rm tot} - Z$:

$$\begin{aligned}\Delta_Z^{(\pm)} &= \frac{4}{\Delta t}\sum_{i,j}|g_{ij}^{(Z)}|^2 \frac{\sin^2(\Delta t(\varepsilon_i^{(Z)} - \varepsilon_j^{(Z)})/(2\hbar))}{(\varepsilon_i^{(Z)} - \varepsilon_i^{(Z)})^2} \\ &\times \ (2J_i+1)(2J_j+1)n_{j,i}^{(Z)}(T)\ (1-n_{i,j}^{(Z)}(T))\end{aligned}\tag{7}$$

IOP Conf. Series: Journal of Physics: Conf. Series **1014** (2018) 012009 doi:10.1088/1742-6596/1014/1/012009

where the matrix elements g_{ij} describe one-nucleon exchange between the nuclei, and their values are calculated microscopically using the expression obtained in Ref. [11]. The occupation numbers of the single-particle states depend on the effective temperature of the nuclei. $\Delta t = 10^{-22}s$ is used in these calculations.

The contribution of the quasifission products to the yield of the ^{36}S+^{238}U reaction products at the beam energy $E_{\text{lab}} = 185$ MeV (has been estimated by the numerical solution of Eq. (4) with the initial conditions $Y_Z(Z_P = 16) = 1$ and $Y_Z(Z_T = 92) = 1$. The single-particle characteristics of the proton and neutron structures of the interacting nuclei are found by the solution of the Schrödinger equation with the Woods-Saxon potential for the set of nuclei $Z = 2$—108 ($A = 4$—274). The last filled level of the single-particle scheme the proton and neutrons are shifted to the values of the proton (S_p) and neutron (S_n) separation energies [12], respectively, to make the the single-particle scheme more realistic. The contribution of the fusion-fission products of the ^{36}S+^{238}U reaction has been estimated by the solution of Eq. (4) with the initial condition $Y_Z(t=0) = \exp(-(Z-Z_0)^2/(2\sigma^2))/(\sqrt(2\pi)\sigma)$, where $Z_0 = (Z_P + Z_T)/2$ and $\sigma = 4$.

The charge distribution of the quasifission products $Y_{\text{qf}}(Z, A)$ is determined by the sum of the DNS decay products during its lifetime t_{int}:

$$Y_{\text{qf}}(Z, A) = \sum_{t=0}^{t=t_f} Y\Lambda_Z^{qf}(E_Z^*))P_Z(E_Z^*, t)\Delta t, \tag{8}$$

where Δt is the step of time in the numerical calculations.

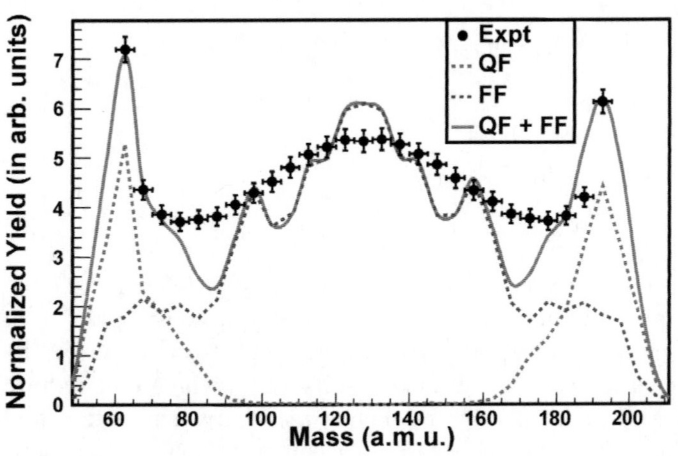

Figure 6. (Color online). The results of the theoretical estimations of the quasifission (red dashed curve) and fusion-fission (blue dashed curve) products in the framework of the DNS model [4, 5] for the ^{48}Ti + ^{208}Pb reaction. The solid curve shows the sum of the both yields. The solid circle represents the normalized experimental mass yield.

This method has been applied to analyze the mass distributions of the fissionlike products observed in the experiment of the ^{48}Ti + ^{208}Pb reaction [15]. Figure 6 represents the comparison of the theoretical predicted mass yield with the experimental ones. Theoretical yield of the reaction products presented in fig. 7 shows that there is the overlap of the mass distributions of the QF and FF products near mass numbers $A \approx 64$. This peak $A \approx 64$ is related to the formation of the isotopes of Fe and Ni with the neutron numbers $N = 38$ and 40.

Theoretical fusion probability is determined by the estimation of capture cross section and competition between complete fusion and quasifission cross sections in the framework of the dinuclear system model [5, 6]. The competition between the complete fusion of nuclei in DNS and quasifission (decay of DNS into two fragments) processes decreases the value of the fusion cross section:

$$\sigma_{\text{fus}}(E_{\text{c.m.}}) = \sum_{\ell=0}^{\ell_d(E_{\text{c.m.}})} (2\ell+1)\sigma_{\text{cap}}(E_{\text{c.m.}}, \ell)P_{\text{CN}}(E_{\text{c.m.}}, \ell). \tag{9}$$

IOP Conf. Series: Journal of Physics: Conf. Series **1014** (2018) 012009 doi:10.1088/1742-6596/1014/1/012009

The maximum value of ℓ leading to capture $\ell_d(E_{c.m.})$ depends on the beam energy and it is calculated by the solution of the radial motion equations (see ref. [5]). Since the capture cross section is equal to the sum of the complete fusion and quasifission cross sections, $\sigma_{\rm cap} = \sigma_{\rm fus} + \sigma_{\rm qfis}$, the quasifission cross section is calculated by the expression

$$\sigma_{\rm qfis}(E_{\rm c.m.}) = \sum_{\ell=0}^{\ell_d}(2\ell+1)\sigma_{\rm cap}(E_{\rm c.m.},\ell)(1-P_{\rm CN}(E_{\rm c.m.},\ell)). \tag{10}$$

It should be stressed that quasifission of dinuclear system can take place at all angular momentum values from $\ell = 0$ to ℓ_d.

The quasifission barrier and fusion probability depend on the angular momentum. Angular distribution of the reaction products is sensitive to the value of the orbital angular momentum since the rotational angle of the DNS formed at the capture is determined by its angular momentum and moment of inertia. This physical quantity was studied by us in Ref. [14].

3. Angular distribution of quasifission products.

The angular distribution of the fusion-fission fragments is expected to be an isotropic while the quasifission products are distributed anisotropically [8]. The lifetime of the fissioning system is an important quantity determining the value of the rotational angle after its formation. The time for formation and fission of compound nucleus is much longer than the lifetime of dinuclear system producing quasifission products. Therefore, in fusion-fission, the fissioning nucleus having angular momentum can make many rotations before fission losing memory about initial orientations and correlation between the fusion-fission fragment masses and angles can disappear.

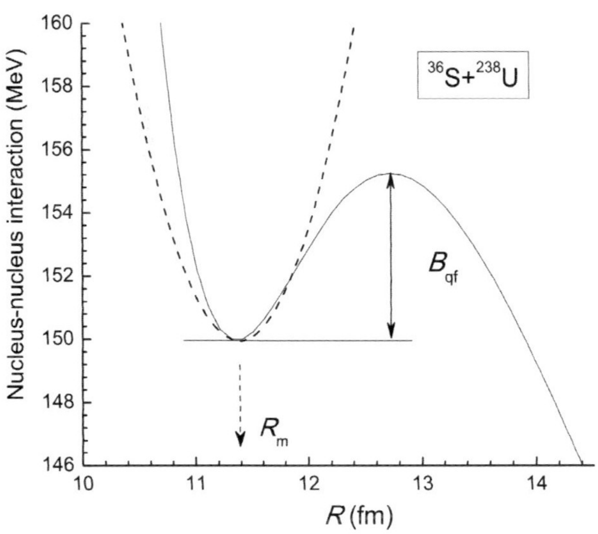

Figure 7. The nucleus-nucleus interaction potential $V(R)$ (solid line) for the ^{36}S + ^{238}U system: the quasifission barrier B_{qf} as the a depth of the potential well. The harmonic oscillator (dashed line) is used to approximate the potential well at $R = R_m$.

The lifetime of dinuclear system is determined by the depth of the potential well $B_{\rm qf}$ (so called quasifission barrier) and amount of its excitation energy $E^*_{\rm DNS}$. The requested decay time τ_Z is estimated by

$$\tau(\ell, T_Z) = \frac{\hbar}{\Gamma^{(Z)}_{qfiss}(T_Z)} \tag{11}$$

if we know the excitation energy E^*_Z and quasifission barrier $B^{(Z)}_{qf}$ of the dinuclear system for its decay on fragments with charge numbers Z and $Z_{tot} - Z$, by using the one-dimensional Kramers rate [16]

$$\begin{aligned}\Gamma^{(Z)}_{qfiss}(\Theta) &= K_{rot}\,\omega_m\left(\sqrt{\gamma^2/(2\mu_{qf})^2+\omega^2_{qf}}-\gamma/(2\mu_{qf})\right)\\ &\times\ \exp\left(-B_{qf}/T_Z)\right)/(2\pi\omega_{qf}).\end{aligned} \tag{12}$$

Here the frequency ω_m and ω_{qf} are found by the harmonic oscillator approximation to the nucleus-nucleus potential $V(R)$ shape for the given DNS configuration $(Z, Z_{tot} - Z)$ on the bottom of its pocket placed at R_m and on the top (quasifission barrier) placed at R_{qf} (see Fig.7), respectively:

$$\omega^2_m = \mu^{-1}_{qf}(R)\left|\frac{\partial^2 V(R)}{\partial R^2}\right|_{R=R_m}, \tag{13}$$

$$\omega^2_{qf} = \mu^{-1}_{qf}(R)\left|\frac{\partial^2 V(R)}{\partial R^2}\right|_{R=R_{qf}}. \tag{14}$$

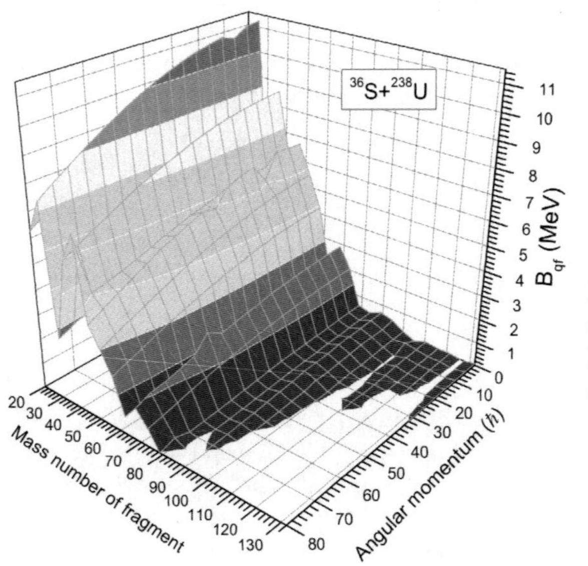

Figure 8. (Color online). The dependence of the quasifission barrier $B_{\rm qf}$ providing the DNS stability on its angular momentum L and mass number one of its fragments.

The calculated values of $\hbar\omega_m$ and $\hbar\omega_{qf}$ were equal to 46.52 MeV and 22.37 MeV, respectively. The used value of the friction coefficient γ is equal to $8 \cdot 10^{-22}$ MeV fm^{-2}s which was found from our calculations; $\mu_{qf} \approx \mu = A_1 \cdot A_2/A_{\rm tot}$, where A_1 and A_2 are the mass numbers of the quasifission fragments.

Depending on the orbital angular momentum L and beam energy $E_{\rm c.m}$ in the entrance channel, the DNS may break up soon after capture, or may stick together for more than one rotation. As long as the projectile-like and target-like nuclei break apart within half a rotation, quasifission events show a strong correlation between mass ratio and angle.

The angular distribution of the quasifission fragments has been theoretically studied by us in Ref. [13]. It was found by estimation of the rotational angle θ_{DNS} during the lifetime $(\tau(T_Z))$ of the system:

$$\theta_Z = \theta_{capture} + \Omega_Z \cdot \tau_Z(\ell, T_Z(E^*_Z(\ell))), \tag{15}$$

where the DNS angular velocity is calculated as ratio of the its angular momentum L and moment of inertia $J^{(Z)}_{\rm DNS}$:

$$\Omega_Z = \frac{L}{J_{\rm DNS}}. \tag{16}$$

$\theta_{capture}$ is the angle between the beam direction and line connecting the mass centres of nuclei at capture. At the given value of L the value of Ω_Z depends only on the DNS moment of inertia at the charge asymmetry Z.

The rotational angle is proportional to the DNS lifetime and during this time the mass distribution between the DNS fragments is changed.

The effective temperature $T_Z(E^*_Z(\ell))$ of the rotating DNS with the charge asymmetry Z is determined by its excitation energy $E^*_Z(\ell)$:

$$T_Z(E^*_Z(\ell)) = \sqrt{\frac{E^*_Z(\ell)}{a}}, \tag{17}$$

where a is the level density parameter taken as $a = A_{\rm tot}/12$ MeV^{-1}. The collective enhancement factor of the rotational motion K_{rot} to the level density should be included because the dinuclear system is a good rotator. It is calculated by the well known expression [17]:

$$K_{rot}(E^*_{DNS}) = \begin{cases} (\sigma^2_\perp - 1)f(E_{DNS}) + 1, & \text{if } \sigma_\perp > 1 \\ 1, & \text{if } \sigma_\perp \le 1, \end{cases}$$

where $\sigma_\perp = J_{(DNS)}T/\hbar^2$; $f(E) = (1 + \exp[(E - E_{cr})/d_{cr}])$; $E_{cr} = 120\widetilde{\beta}^2_2 A^{1/3}$ MeV; $d_{cr} = 1400\widetilde{\beta}^2_2 A^{2/3}$. $\widetilde{\beta}$ is the effective quadrupole deformation for the dinuclear system. We find it from the calculated $\mathcal{J}^{(DNS)}_\perp$.

The mass distribution evolution as diffusion process is determined by the DNS lifetime and shell effects in the nucleon structure in the interacting nuclei. At the same time the DNS lifetime depends on its mass (charge) asymmetry since the quasifission barrier $B_{\rm qf}$ is found as the depth of the potential well. At capture of the projectile-nucleus by the target-nucleus the formed DNS acquires angular momentum and it rotates around own centre-of-mass.

It should be noted the evolution of the mass distribution between the DNS fragments and its rotation around the centre-of-mass system take place simultaneously. In other words the during rotation of DNS its momentum inertia is changed and its lifetime depends on the mass asymmetry since the quasifission barrier is related on the charge and mass asymmetry. It is clear that the more mass asymmetric configurations of DNS rotates longer than the less mass asymmetric configurations. Certainly the DNS configurations having large population due to shell effects or/and initial charge and mass numbers of projectile- and target- nuclei can show intense yield of the reaction products. But large quasifission barrier causes hindrance decay from the mass asymmetric configurations and DNS has more chance for decay from less mass asymmetric configurations. The possibility to reach the more symmetric configuration depends on the DNS excitation energy E^*_{DNS} and peculiarities of the driving potential. The both of these last quantities depends on the DNS angular momentum L. Therefore, simultaneous calculations of the mass and angular distributions as a function of the beam energy $E_{\rm c.m.}$ and orbital angular momentum are important for the study of the reaction mechanism producing quasifission products.

The left panel of Fig. 4 and Fig. 5 show that the mass distributions of the quasifission products are sufficient in the mass number ranges $60 \div 90$ and $180 \div 210$. We can conclude that the lifetime of the rotating DNS is enough for evolution of the mass distribution from the initial values $A_P = 36$ and $A_T = 238$ into the corresponding mass ranges. The maximum of the mass distribution of the quasifission products is related with the formation of the ^{68,70}Ni

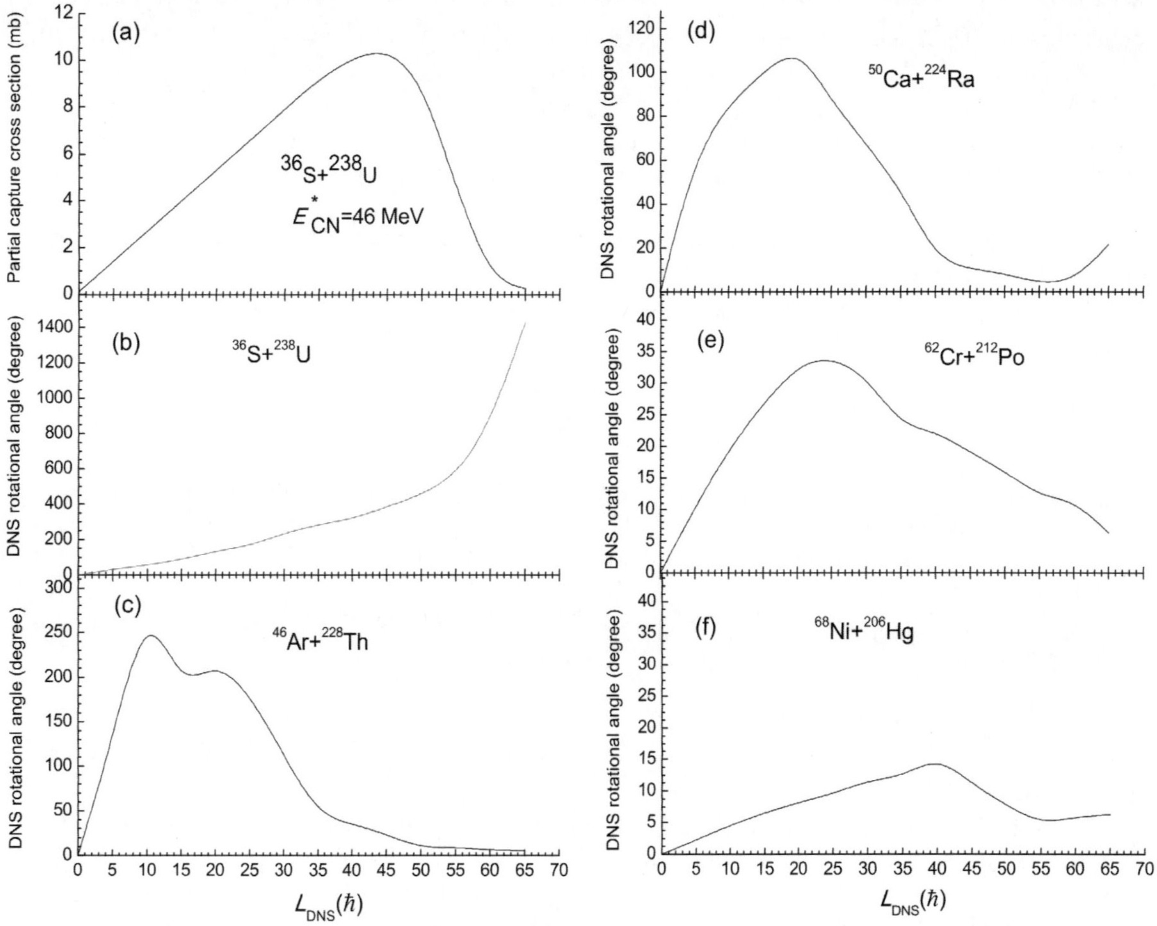

Figure 9. (Color online). The angular momentum distribution calculated for the DNS formed at capture in the ^{36}S+^{238}U reaction at $E_{\rm lab}$= 185 MeV (corresponding excitation is $E^*_{\rm CN}$= 46 MeV)(a). Rotational angles for the DNS calculated for its configurations consisting from the interacting fragments ^{36}S+^{238}U (b), ^{46}Ar+^{228}Th (c), ^{50}Ca+^{224}Rn (d), ^{62}Cr+^{212}Po (e) and ^{64}Ni+^{210}Hg (f).

isotopes. The conjugate heavy fragments are Hg isotopes with the magic neutron numbers around $N \approx 126$. While the mass distribution is changed the DNS is rotating. The dependence of the rotating angles on the mass and charge numbers of the interacting nuclei is presented in Fig. 9 for the pairs of nuclei ^{36}S+^{238}U (9b), ^{46}Ar+^{228}Th (9c), ^{50}Ca+^{224}Rn (9d), ^{62}Cr+^{212}Po (9e) and ^{64}Ni+^{210}Hg (9f). The partial capture cross section for the ^{36}S+^{238}U reaction at $E_{\rm lab}$= 185 MeV (corresponding excitation is $E^*_{\rm CN}$= 46 MeV) is presented in Fig. 9a. One can say that population of the DNS angular momentum is different for various reactions and beam energies.

The fusion probability is ^{36}S+^{238}U is about $P_{\rm CN} = 0.3$, therefore, the contribution of the fusion-fission products are comparable with the one of the quasifission products as seen from the right panel of Fig. 4. In case of the ^{48}Ca+^{238}U reaction, the fusion probability is much smaller $P_{\rm CN} = 0.1$ and the yield of the fusion-fission products is smaller than the yield of quasifission products. We conclude that the study of the mass and angular distribution of the reaction products allows us to estimate fusion probability. The uncertainties in estimation of the fusion probability from the analysis of the fission-like products is related with separation of the

IOP Conf. Series: Journal of Physics: Conf. Series **1014** (2018) 012009 doi:10.1088/1742-6596/1014/1/012009

contributions of quasifission and fusion-fission products when there is a doubt of their overlap.

References

[1] Itkis M G *et al* 2006 *AIP Conference Proceedings* **853** 231
[2] Itkis M G *et al* 2007 *Nucl. Phys. A* **787** 150c
[3] Itkis M G *et al* 2015 *Nucl. Phys. A* **944** 204
[4] Fazio G *et al.* 2005 *Mod. Phys. Lett. A* **20** 391
[5] Nasirov A K *et al.* 2005 *Nucl. Phys. A* **759** 342
[6] Fazio G *et al.* 2005 *Phys. Rev. C* **72** 064614
[7] Sierk A J 1986 *Phys. Rev. C* **33** 2039
[8] Shen W Q *et al.* 1987 *Phys. Rev. C* **36** 115
[9] Nasirov A K *et al.* 2013 *Eur. Phys. J. A* **49** 147
[10] Jolos R V, Muminov A I and Nasirov A K 1986 *Yad. Fiz.* **44** 357; 1986 *Sov. J. Nucl. Phys.* **44** 228
[11] Adamian G G, Jolos R V and Nasirov A K 1992 *Yad. Fiz.* **55** 660, 1992 *Sov. J. Nucl. Phys.* **55** 366
[12] Möller P, Nix N R, Myers W D and Swiatecki W J 1995 *Atomic Data and Nuclear Data Tables* **59** 185
[13] Nasirov A K *et al.* 2007 *Eur. Phys. J. A* **34** 325
[14] Nasirov A K *et al.* 2009 Phys.Rev. C 79 024606
[15] Meenu Thakur it et al. 2017 *Eur. Phys. J. A* **53** 133
[16] Kramers H A 1940 *Physica* **7** 284; Strutinsky V M 1973 *Phys. Lett. B* **47** 121
[17] Junghans A R *et al.* 1998 *Nucl.Phys. A* **629** 635

ALICE results from Run-1 and Run-2 and perspectives for Run-3 and Run-4

Francesco Noferini on behalf of the ALICE Collaboration
INFN, Sezione di Bologna, Italy
E-mail: Francesco.Noferini@bo.infn.it

Abstract. A review of ALICE is presented focusing on its physics programme and results from the Run-1 and Run-2 data taking periods. Among the four major LHC experiments, ALICE is devoted to the study of the Quark-Gluon Plasma produced in ultra-relativistic heavy-ion collisions (Pb–Pb), but it is also collecting data in smaller systems (pp and p–Pb). This review focuses on the main results collected so far, including the characterization of the QGP via soft and hard probes, and the production rate of light nuclei and hypernuclei. Finally, the perspectives after the detectors upgrades to be performed during 2019-2020 are presented.

1. Introduction

The ALICE detector at the LHC [1, 2] has been conceived specifically to study heavy-ion physics, with a precise tracking of the very high-multiplicity events expected in such collisions and with a powerful particle identification based on different techniques, applicable in a wide momentum range. These features are complementary to the other LHC experiments more focused on hard physics in pp collisions.

The detector (Fig. 1) consists of a central region with tracking and PID detectors embedded in a solenoid with magnetic field B=0.5 T, and a forward arm for muon measurements down to pseudorapidity $\eta \sim 4$. ALICE collected Pb–Pb, pp and p–Pb collisions during LHC Run-1 and Run-2 as reported in Tab. 1.

Differently from hadrons, heavy ions are much more extended objects and therefore the characteristics of a collision depend strongly on the relative distance in the transverse plane between the two colliding nuclei. Usually a nucleus–nucleus collision can be seen as an overlap of independent nucleon–nucleon collisions: the number of such collisions depends on the impact parameter which is the distance between the centers of the two colliding nuclei in the plane orthogonal to the beam direction. However such a parameter is not a direct observable in heavy-ion collisions and an experimental definition is usually based on the multiplicity of charged particles, on the transverse energy at midrapidity or on the energy measured in the forward rapidity region. In particular, measurements in ALICE are mostly based on the amplitude of the V0-scintillator signal [3], whose distribution is fitted using a Glauber model [4] to define the centrality classes. Indeed events are grouped into different classes (*centralities*) which represent the fraction, in percent, of minimum bias collisions starting from those with small impact parameter (centrality $\sim 0\%$) to the most peripheral events ($\sim 100\%$). The Glauber model also allows us to correlate the centrality with some basic parameters of the collision: the number of participating nucleons ($N_{\rm part}$) and the number of binary collisions ($N_{\rm coll}$).

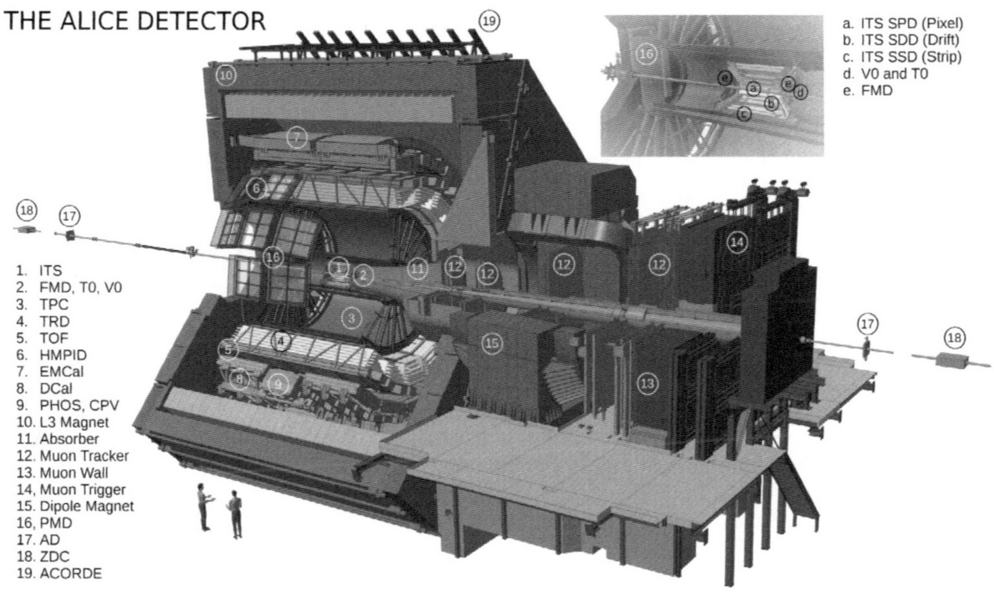

Figure 1. The ALICE detector in Run-2 at the LHC.

Year	Systems	$\sqrt{s_{\rm NN/pN}}$ (TeV)	Integrated Luminosity ($\rm nb^{-1}$)
Run-1			
2009-2010	pp	0.9	$\sim$ 0.15
2011	pp	2.76	$\sim$ 1.1
2010-2011	pp	7	$\sim$ 4800
2012	pp	8	$\sim$ 9700
2013	p–Pb	5.02	$\sim$ 30
2010-2011	Pb–Pb	2.76	$\sim$ 0.1
Run-2			
2015	pp	5.02	$\sim$ 100
2015-2016	pp	13	$\sim$ 14000
2016	p–Pb	5.02	$\sim$ 3
2016	p–Pb	8.16	$\sim$ 20
2016	Pb–p	8.16	$\sim$ 20
2015	Pb–Pb	5.02	$\sim$ 0.4
2017	Xe–Xe (B=0.2 T)	5.44	to be estimated

Table 1. ALICE data taking periods. The nominal magnetic field was B = 0.5 T except for Xe–Xe system and few pp data samples at $\sqrt{s}$ = 13 TeV in 2015-2017.

The evolution of the system produced in ion-ion collisions can be schematically described as following: the Lorentz-contracted colliding nuclei produce, in a time $\leq$1 fm/c, a hot system of quarks and gluons interacting inelastically, which later thermalizes into a deconfined phase lasting of the order of a few fm/c and containing all allowed $\rm SU(3)_{color}$ states. The system then expands and cools down until particles with zero color are formed and a chemical freezout temperature is reached. At this moment the particle composition is fixed. Further expansion and

IOP Conf. Series: Journal of Physics: Conf. Series **1014** (2018) 012010 doi:10.1088/1742-6596/1014/1/012010

cooling brings the system to a kinematic freezout temperature when no more elastic collisions are possible and the kinematic properties of the particles are defined. While electroweak interacting particles (photons, Z, W, ...) pass the medium unchanged, partons are strongly affected by the dense/hot matter. For this reason hadron production at very high transverse momentum (p_T) is a very effective observable to explore energy loss via QCD interactions in a highly dense medium.

In the next sections a selection of ALICE results is presented (a more detailed review can be found in [5]).

2. Soft probes

Soft probes (low p_T observables) are sensitive to the full evolution of the medium produced in ultra-relativistic heavy-ion collisions and therefore are very important in characterizing the properties of the QGP (Quark Gluon Plasma) in terms of mechanisms for hadron production, statistical/thermal features and collective phenomena developed in such a regime.

2.1. HBT correlations

The fireball spatial extent at decoupling is accessible via interferometry, a technique which exploits the Bose-Einstein enhancement of identical bosons close in phase space. In nuclear physics the strength of such a correlation for identical bosons is sensitive to different geometrical parameters (radii) of the emitting source. The size of the medium at freeze-out can be characterized by the correlators along the three axes providing the corresponding radii R_{out} and R_{side} (the size in the transverse plane) and R_{long} (the size along the longitudinal direction). The estimate of the fireball volume relies on the product of the three radii. In Fig. 2 the volume of the system is reported for different center-of-mass energies as a function of the average charged-track multiplicity density, $\langle dN_{\text{ch}}/d\eta\rangle$. The ALICE results give a source volume at the LHC energies $V \sim 300$ fm^3, a factor 2 larger than the volume measured at RHIC, while the corresponding lifetime is estimated to be $\tau \sim 10$ fm/c, a 20% longer than that measured at RHIC.

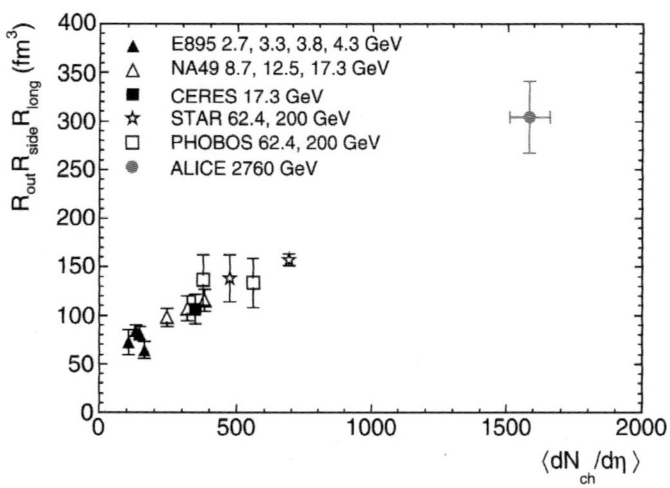

Figure 2. Volume of the fireball [6] vs. charged-track multiplicity density from two-pion HBT correlation: ALICE results compared to lower energy experiments.

2.2. Hadron production

Other observables in the soft sector (mass, meson/baryon ratio, ...) are based on the outstanding capabilities of ALICE for particle identification. Many features of the system produced in heavy-ion collisions are strongly dependent on the different particle properties, and consequently the ALICE detector was optimized to identify hadrons in a very wide p_{T} range, from very low p_{T} ($\sim$ 150 MeV/c) to p_{T} as large as few tens of GeV/c.

Hadron production was measured at the LHC in central and peripheral Pb–Pb collisions [7, 8]. The p_{T} dependence of pion, kaon and proton spectra was found to be much harder than at RHIC, as predicted by models based on the hydrodynamic expansion of the produced medium (Fig. 3, left). The centrality dependence of the p/π ratio production shows the role of the particle masses in such expansion. In the most central events, protons appear to be pushed to higher momenta under a higher pressure gradient (Fig. 3, right): this is a typical feature of a hydrodynamical mechanism in the presence of a collective motion.

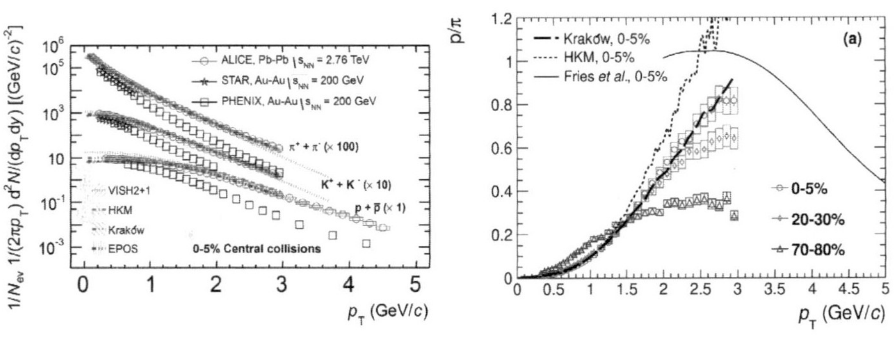

Figure 3. Left: p_{T} spectra of particles for summed charged states at 0-5% centrality, compared to hydrodynamical models and results from RHIC. Right: Proton over pion ratio as a function of p_{T} at different centralities. Both figures are taken from [8].

In addition, the enhancement in baryon production with respect to mesons named "baryon anomaly" [9, 10], is pronounced also in the strangeness sector, as for the Λ/K ratio [11] (Fig. 4). This effect, strongly centrality dependent, may indicate not only an increase of radial flow but also a change in the production mechanism, with hadron production by hadronization being coupled to quark recombination. Production via recombination is expected to play an important role at intermediate p_{T}, when partons start to be closer in phase space due to an increase of their density. To distinguish effects related to the mass or to the quark content a comparison between protons and ϕ-mesons is very useful. The flatness of the ϕ/p ratio reported in Fig. 4 may indicate that the production of hadrons for $p_{\mathrm{T}} < 2$ MeV/c is driven by their mass.

In general, strangeness production is expected to be favored in Pb–Pb collisions as observed at AGS [12], SPS [13, 14] and RHIC [15] if a deconfined phase is created. Multistrange production is a good obvservable to quantify it: in ALICE this measurement [16] was performed following the same prescription adopted by previous experiments (Fig. 5). At the LHC energies an enhancement of the Ξ and Ω production was found when moving from pp to Pb–Pb collision systems, similarly as observed at lower energies. The predictions for Ξ/π and Ω/π at the LHC from the thermal models are also shown; they are based on a grand canonical approach, described in [17] (full line, with a chemical freeze-out temperature parameter T = 164 MeV) and [18] (dashed line, with T = 170 MeV).

More recently, ALICE measured the production of strange baryons also for p–Pb in different multiplicity classes (Fig. 6) [19]. The plateau in the multistrange-over-pion ratio for high

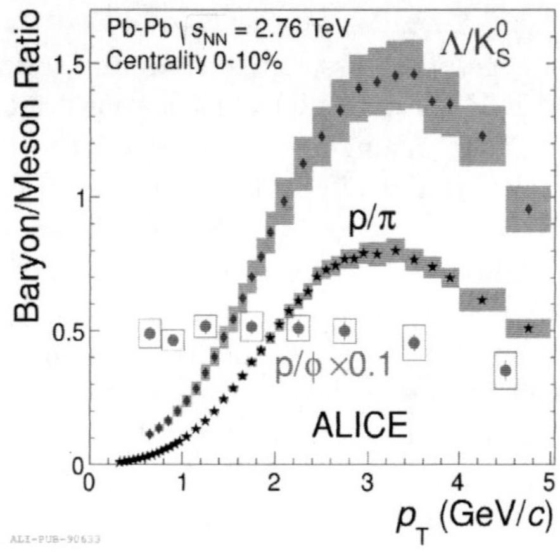

Figure 4. Λ/K^0_s vs. p_T compared to the p/π ratio [11]. The ϕ/p ratio is reported as well.

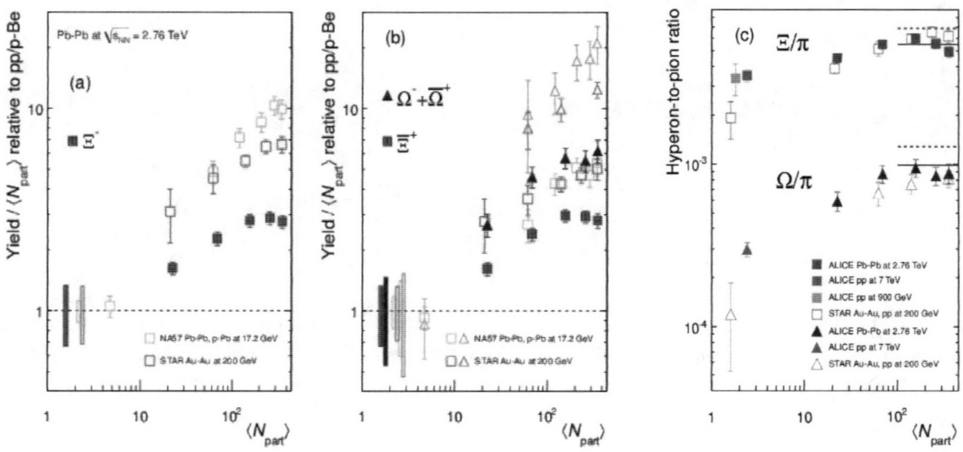

Figure 5. Strangeness enhancements in the rapidity range $|y| < 0.5$ as a function of the mean number of participants, N_{part}, showing LHC (ALICE, full symbols), RHIC and SPS (open symbols) data [16].

multiplicity events is consistent with the prediction of some thermal models which expect a saturation to the grand canonical value when the size of the system is larger than several fm. The rise in the low/intermediate multiplicty region may indicate that such a saturation is not still reached.

2.3. Anisotripic flow

ALICE studied in details the anisotropy in momentum space of produced particles. If the matter is strongly interacting, intial anisotropies in coordinate space, coming either from the geometry of the collision or from fluctuations, convert into an anisotropic momentum distribution. An illustration of the anisotropy development in a strongly interacting medium is reported in Fig. 7.

In order to reconstruct this anisotropy the particle distribution can be decomposed, event by

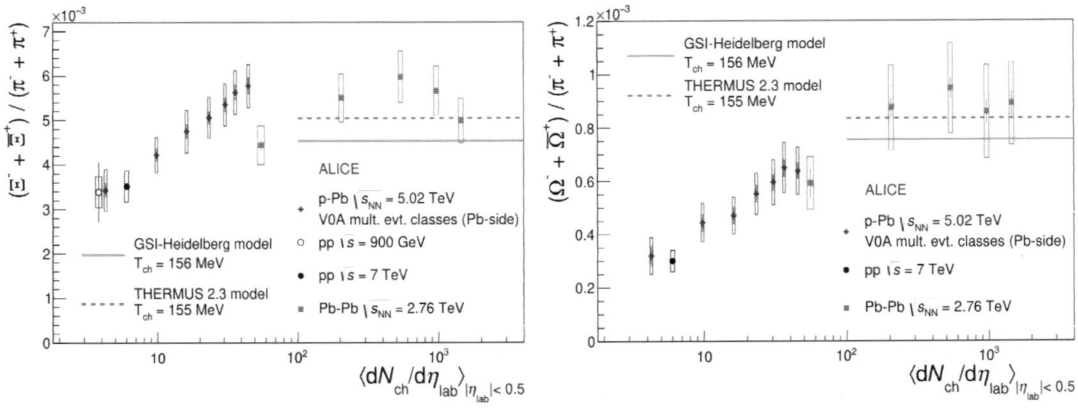

Figure 6. Ξ and Ω over π ratios vs. charged track multiplicity density in pp, p–Pb and Pb–Pb collisions [19].

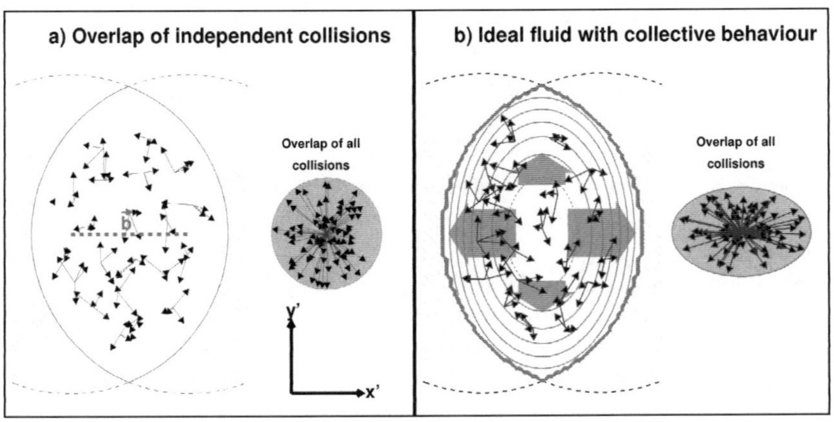

Figure 7. Due to an initial spatial anisotropy of the region where nucleons interact, different pressure gradients may develop along the in-plane and out-of-plane directions in a strongly interacting medium. Therefore, despite of an isotropic momentum distribution of the initial states, the momenta of the final products are anisotropically distributed. The different scenarios, with (right) and without (left) the assumption of a strongly interacting matter, are shown.

event, in a Fourier expansion:

$$dN/d\phi \propto 1 + \sum_{n=1} 2v_n \cos\left(n\left(\phi - \Psi_n\right)\right), \tag{1}$$

where ϕ is the azimuthal angle of the produced particles, Ψ_n the reaction plane angle of each harmonic and v_n the coefficients of the anisotropy magnitudes.

In order to clarify the role of radial expansion in the development of v_2 (elliptic flow), the mesurement has to be performed separately for each identified particle [20]. For instance, at low transverse momenta the ϕ-meson elliptic flow is similar to that of the (anti)proton: this can be explained as a similar effect of radial flow on particles with similar mass. At intermediate transverse momenta, $p_{\rm T} > 2.5$ GeV/c, the hydrodinamic regime does not provide a good description. In such a regime other processes, like quark recombination, may play a more important role leading to a different behavior depending on the quark content (meson/baryon

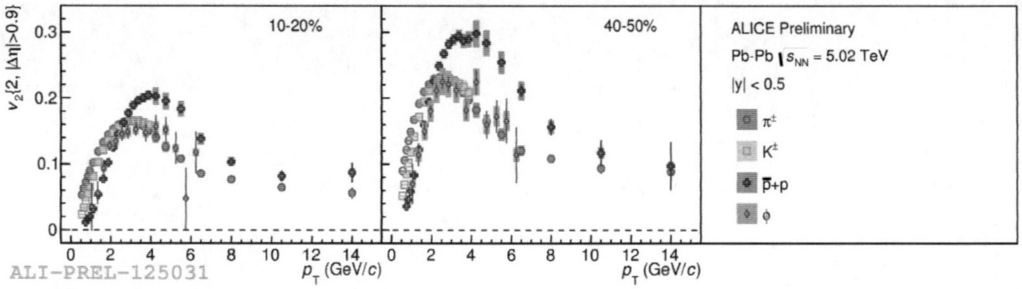

Figure 8. The p_T-differential v_2 for different particle species [20], measured in Pb–Pb collisions at $\sqrt{s_{NN}} = 5.02$ TeV. The left and right plots present the comparison for the 10-20% and 40-50% centrality intervals, respectively.

splitting). In particular, the ϕ meson "stays" with other mesons above 3 GeV/c accordingly to the prediction of coalescence/recombination models.

3. Hard probes

Hard probes have been investigated deeply in the previous energy regimes [22]. It is generally accepted that, before hadronization, partons lose energy in the medium via gluonstrahlung or in multiple collisions [23, 24, 25]. This process is called "quenching", was predicted by QCD [26] and was discovered at RHIC [27].

Several observables have been proposed to probe the hard-parton interactions with the medium. One of the most direct ones used to characterize the energy-loss mechanism is the nuclear modification factor R_{AA}, which quantifies the change in particle p_T distributions from pp to Pb–Pb systems:

$$R_{AA}(p_T) = \frac{d^2N^{AA}/d\eta dp_T}{\langle T_{AA}\rangle \, d\sigma^{pp}/d\eta dp_T}, \tag{2}$$

where T_{AA} is the nuclear-overlap function derived with a Glauber calculation [4], which accounts for the average number of binary collisions occurring in AA systems for a given centrality selection. If heavy-ion interactions are just a mere overlap of N_{coll} binary nucleon–nucleon collisions, then R_{AA} is expected to be one.

As shown in Fig. 9, hadrons are strongly suppressed, while pure electroweak probes are compatible with unity.

3.1. Charmed probes

Charm and beauty quarks are produced in the early stage of the collision. Their production in pp collisions at the LHC is a tool to test pQCD calculations in a new energy domain. As for charged particles the nuclear modification factor was computed in several centrality selections [29]. The results are reported in Fig. 10: the suppression for charm mesons is very clear, indicating a strong interaction with the hot medium produced. The strength of R_{AA} suppression is almost as large as that observed for charged particles, which are mainly light-flavor hadrons, with a possible indication of a slightly higher value for open-charm states in the lower p_T region.

As for hadrons and jets, the same measurement was performed in p–Pb collisions for centrally produced D mesons [30]. Within the errors, the data do not show any sign of suppression, compatible with the absence of cold-nuclear effects and the genuine suppression in Pb–Pb collisions due to the interaction of the heavy quarks with the medium produced.

IOP Conf. Series: Journal of Physics: Conf. Series **1014** (2018) 012010 doi:10.1088/1742-6596/1014/1/012010

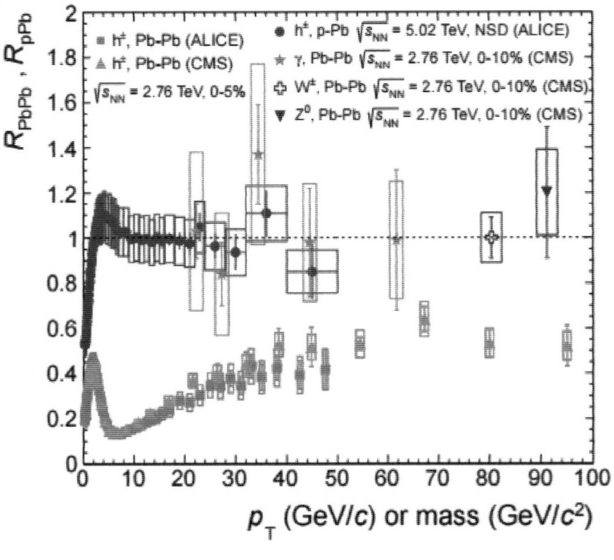

Figure 9. Nuclear modification factor R_{AA} of hadrons measured by ALICE in p–Pb and Pb–Pb collisions [28].

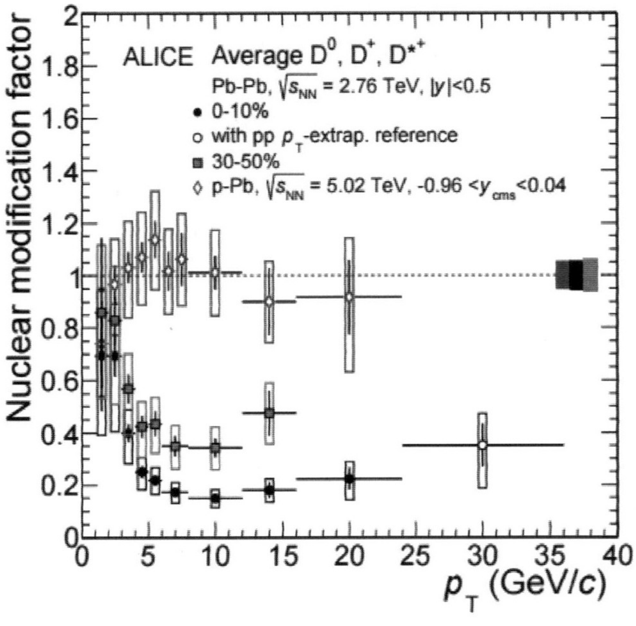

Figure 10. Average R_{AA} for prompt D^0, D^+ and D^{*+} in p–Pb and Pb–Pb 0–10%(30–50%) versus p_{T}.

According to the color-screening model [31], the measurement of the in-medium dissociation probability of the different quarkonium states is expected to provide an estimate of the initial temperature of the system.

Several models attempt to reproduce the RHIC data by adding to the direct J/Ψ production a regeneration component from deconfined charm quarks in the medium. This charmonium regeneration is stronger at the LHC where the high-energy density of the medium and the large number of $c\bar{c}$ pairs produced in central Pb–Pb collisions favor this extra production.

The nuclear modification factor for inclusive J/Ψ production is reported in Fig. 11 [32, 33, 34].

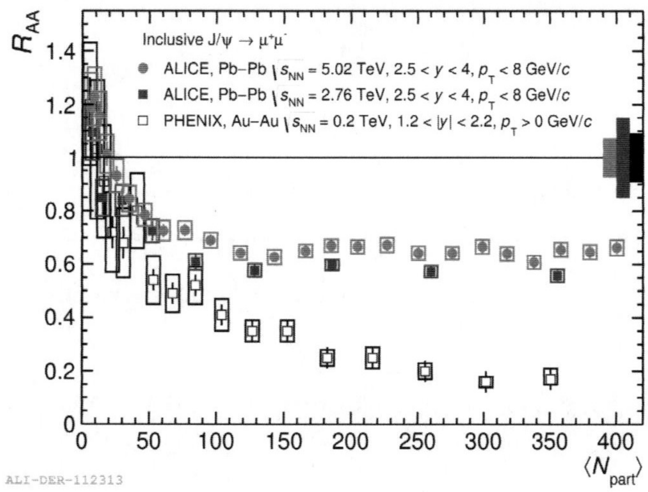

Figure 11. Inclusive J/Ψ $R_{\rm AA}$ [33, 34] as a function of the number of participating nucleons as measured in Pb–Pb collisions at $\sqrt{s_{\rm NN}} = 2.76$ TeV, compared to PHENIX results [35, 36] in Au–Au collisions at $\sqrt{s_{\rm NN}} = 200$ GeV at forward rapidity.

The ALICE measurement is compared to PHENIX results [35, 36] (Fig. 11). Focusing on central collisions, at forward rapidity ALICE results are larger by a factor 3 than the PHENIX ones. This is consistent with a larger J/Ψ regeneration at the LHC.

3.2. Collectivity of charm quarks

If the charm quark shares the collective motion of the medium with the light quarks it should be visible looking at observables strongly correlated with the bulk properties. For this reason it is very interesting to look for non-zero correlation in the anosotropic coefficients (flow). ALICE measured v_2 for D mesons in semi-central collisions. D mesons v_2 is positive in the range $2 < p_{\rm T} < 6$ GeV/c with 5.7σ significance [37] and comparable in magnitude to that of light flavor hadrons, as shown in Fig. 12-left.

For semi-central collisions, indications of a non zero J/Ψ v_2 is also observed in the intermediate $p_{\rm T}$ range, as shown in Fig. 12-right [38].

Both results seem in agreement with the global picture in which a significant fraction ($\sim 30\%$ according to the models reported in the figure) of the observed J/Ψ yield is coming from recombination of charm quarks in the deconfined phase.

4. (Anti)nuclei and Hypernuclei production

Ultra-relativistic ion collisions create suitable conditions for producing light (anti)nuclei, because a high-energy density is reached over a large volume. In relativistic heavy-ion collisions, nuclei and corresponding antinuclei are produced with nearly identical rates. In Fig. 13-left the production rate for nuclei and antinuclei is reported. When increasing the atomic number the

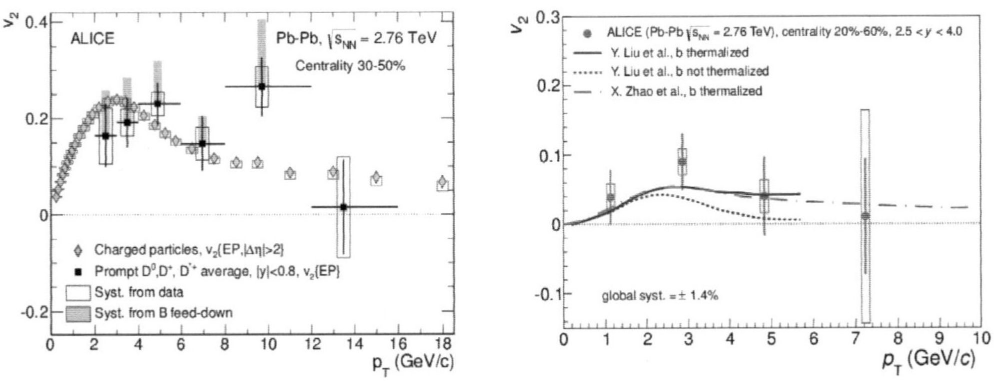

Figure 12. Left: Average of D^0, D^+ and D^+ v_2 as a function of p_T [37], compared to charged particle v_2 [21] in non central (30-50%) Pb–Pb collisions. Right: Inclusive J/Ψ $v_2(p_T)$ for non central (20-60%) Pb–Pb collisions [38]. Calculations from two transport models are also shown.

production is suppressed nearly exponentially accordingly to the expectation both from thermal and coalescence models [39]. In the coalescence picture the production rate for a nucleus with atomic number A depends on the baryon density (nuclei are formed by coalescence of baryons) and it may be approximated by the formula:

$$E_A \frac{\mathrm{d}N_A}{\mathrm{d}p_A^3} = B_A \left(E_\mathrm{p} \frac{\mathrm{d}N_\mathrm{p}}{\mathrm{d}p_\mathrm{p}^3} \right)^A , \tag{3}$$

where E and p are, respectively, the energy and momentum of particles, A(p) refer to nucleus (proton) and B_A is the coalescence parameter, namely the phase-space volume allowed for coalescence. It is worth to note that Eq. 3 is an approximation because it assumes that the volume of the source is constant and then re-assorbed in the B_A parameter. ALICE measured the B_A parameter in different collision systems (accordingly to Eq. 3) and the results are reported in Fig. 13-right. As it can be noted, when increasing the multiplicity of the system the coealescence parameter decreases. This could be ascribed to the increasing volume of the source emitting the nuclei which is not taken into account in this approximation.

In any case, such a measurment is really important for astroparticle experiments to estimate the possibility to create antinuclei by coalescence of antibaryons. In particular the AMS experiment [41] can benefit from the ALICE measurement of B_3 [42] to reduce the uncertanties of the background contamination when looking at the primary $^3\overline{\mathrm{He}}$ detection.

In addition to the study of (anti)nuclei ALICE is also able to reconstruct the daughter particles emitted in hypernuclei decay. $^3_\Lambda\mathrm{H} \rightarrow ^3\mathrm{He} + \pi^-$ (and charge conjugated, branching ratio $\sim 25\%$) is an efficient channel to detect hypernuclei. In Fig. 14-left a clear peak in the invariant mass distribution for the channel considered is visible. By reconstructing the secondary vertices of the decays ALICE measured the lifetime of this state with the best precision reached so far (Fig. 14-right).

Such measurements allow us to understand better the baryon-baryon interactions in the strangeness sector and to study the structure of multistrange systems.

5. The ALICE upgrade for Run-3 and Run-4

The LHC machine will provide in the next years Pb–Pb collisions with increasing luminosity. The current schedule is reported below:

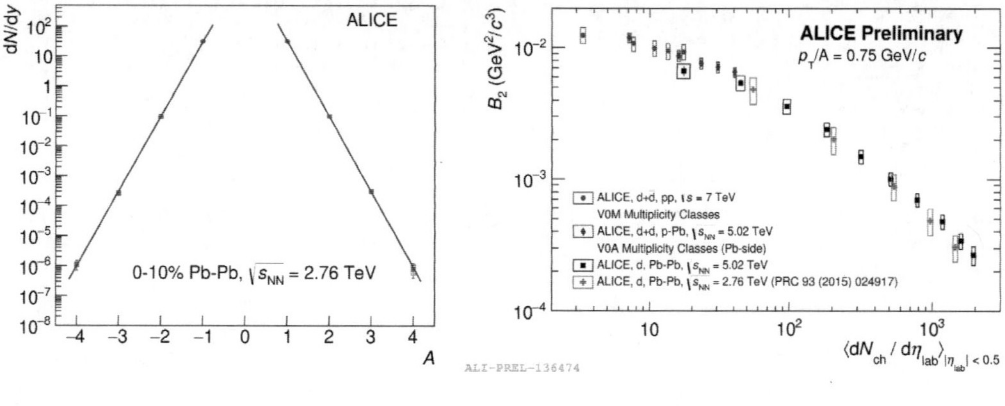

Figure 13. Left: dN/dy for protons (A=1) up to ^{4}He (A=4) and the corresponding anti-particles in central (0-10%) Pb–Pb collisions at $\sqrt{s_{\rm NN}} = 2.76$ TeV [39]. Right: B_A parameter in different collision systems as a function of charged particle multiplicity [40].

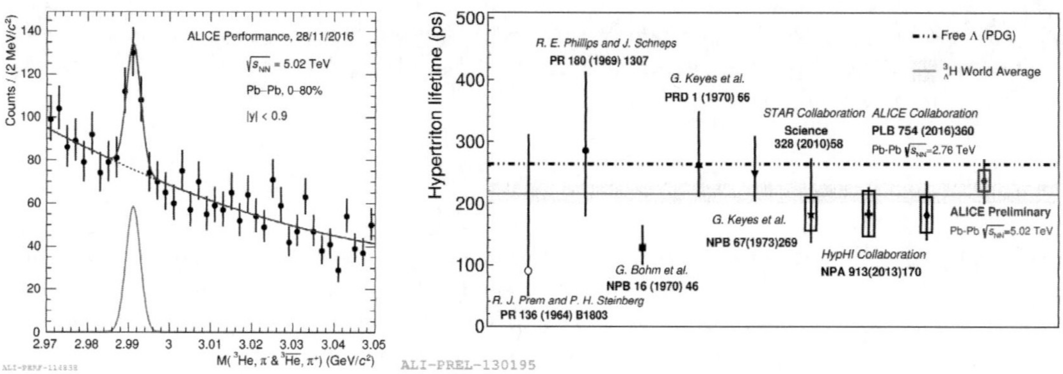

Figure 14. Left: $^3{\rm He} + \pi$ invariant mass distribution in Pb–Pb at $\sqrt{s_{\rm NN}} = 5.02$ TeV. Right: Hypertriton lifetime.

- Run-2 (ongoing up to 2018) collecting a luminosity in Pb–Pb of 1 nb^{-1};
- Run-3 (2021-2023) expecting a luminosity in Pb–Pb of 6 nb^{-1};
- Run-4 (2026-2029) expecting a luminosity in Pb–Pb of 7 nb^{-1}.

In order to benefit from the increasing statistics ALICE is going to upgrade several detector systems during the Long Shutdown 2 (2019-2020). One of the major changes is the upgrade to faster detectors and continuous readout to increase the acquisition rate of the events [43]. In this respect the major improvement concerns the time-projection chamber which will be upgraded with a GEM technology [44]. On parallel the online and offline systems will be completely renewed [45] to allow us to manage the new features of the DAQ model. One of the main goals in Run-3 and Run-4 is the high-precision measurement of charm and beauty quark properties in the medium. This will be achieved by improving the capability of the secondary-vertex reconstruction after the ITS upgrade [46] and the installation of the Muon Forward Tracker

[47]. To extend the study of the thermal properites of the medium using thermal photons and low-mass dileptons emitted by the QGP, some dedicated runs will be taken at a lower magnetic field (B=0.2 T). This specific configuration was already validated during Run-2 with Xe–Xe collisions, but a large data sample with such a condition will be reached only at Run-3.

References

[1] ALICE Collaboration (Aamodt K *et al*) 2008 *JINST* **3** S08002
[2] ALICE Collaboration (Abelev B *et al*) *Int. J. Mod. Phys. A* 2014 **29** 1430044
[3] ALICE Collaboration (Abbas E *et al*) *JINST* 2013 **8** P10016
[4] ALICE Collaboration (Abelev B *et al*) 2013 *Phys. Rev. C* **88** 044909
[5] Cifarelli L, Nania R, Noferini F and Zichichi A 2016 *Nuovo Cimento* **39** 10
[6] ALICE Collaboration (Aamodt K *et al*) 2011 *Phys. Lett. B* **696** 328
[7] ALICE Collaboration (Abelev B *et al*) 2012 *Phys. Rev. Lett.* **109** 252301
[8] ALICE Collaboration (Abelev B *et al*) 2013 *Phys. Rev. C* **88** 044910
[9] Vitev I and Gyulassy M 2002 *Phys. Rev. C* **65** 041902
[10] PHENIX Collaboration (Adcox K *et al*) 2002 *Phys. Rev. Lett.* **88** 242301
[11] ALICE Collaboration (Abelev B *et al*) 2013 *Phys. Rev. Lett.* **111** 222301
[12] E-802 Collaboration (Ahle L *et al*) 1998 *Phys. Rev. C* **58** 3253
[13] NA57 Collaboration (Antinori F *et al*) 2006 *J. Phys. G* **32** 427
[14] NA57 Collaboration (Antinori F *et al*) 2010 *J. Phys. G* **37** 045105
[15] STAR Collaboration (Abelev B *et al*) 2008 *Phys. Rev. C* **77** 044908.
[16] ALICE Collaboration (Abelev B *et al*) 2014 *Phys. Lett. B* **728** 216
[17] Andronic A, Braun-Munzinger P and Stachel J 2009 *Phys. Lett. B* **673** 142
[18] Cleymans J *et al* 2006 *Phys. Rev. C* **74** 034903
[19] ALICE Collaboration (Adam J *et al*) 2016 *Phys. Lett. B* **758** 389
[20] ALICE Collaboration (Bertens R A *et al*) 2017 *Nucl. Phys. A* **967** 385
[21] ALICE Collaboration (Aamodt K *et al*) 2010 *Phys. Rev. Lett.* **105** 252302.
[22] D. d'Enterria, "Nuclei–Relativistic Heavy Ion Physics" edited by Landolt-Bornstein (Springer-Verlag, Berlin) 2010, Group 1 Volume 23.
[23] J. D. Bjorken, FERMILAB-PUB-82-059-THY, FERMILAB-PUB-82-059-T (1982).
[24] Gyulassy M and Plumer M 1990 *Phys. Lett. B* **243** 432
[25] Wang X and Gyulassy M 1992 *Phys. Rev. Lett.* **68** 1480
[26] Baier R, Schiff D and Zakharov B G 2000 *Ann. Rev. Nucl. Part. Sci.* **50** 37
[27] PHENIX Collaboration (Adcox K *et al*) 2002 *Phys. Rev. Lett.* **88** 022301
[28] ALICE Collaboration (Abelev B *et al*) 2013 *Phys. Rev. Lett.* **110** 082302
[29] ALICE Collaboration (Adam J *et al*) 2016 *JHEP* **1603** 081
[30] ALICE Collaboration (Abelev B *et al*) 2014 *Phys. Rev. Lett.* **113** 232301
[31] Matsui T and Satz H 1986 *Phys. Lett. B* **178** 416
[32] ALICE Collaboration (Abelev B *et al*) 2012 *Phys. Rev. Lett.* **109** 072301
[33] ALICE Collaboration (Adam J*et al*) 2016 *JHEP* **1605** 179
[34] ALICE Collaboration (Adam J *et al*) 2017 *Phys. Lett. B* **766** 212
[35] PHENIX Collaboration (Adare A *et al*) 2007 *Phys. Rev. Lett.* **98** 232301
[36] PHENIX Collaboration (Adare A *et al*) 2011 *Phys. Rev. C* **84** 054912
[37] ALICE Collaboration (Abelev B *et al*) 2013 *Phys. Rev. Lett.* **111** 102301
[38] ALICE Collaboration (Abbas E *et al*) 2013 *Phys. Rev. Lett.* **111** 162301
[39] ALICE Collaboration (Acharya S *et al*) 2018 *Nucl. Phys. A* **971** 1
[40] ALICE Collaboration (Adam J *et al*) 2016 *Phys. Rev.C* **93** 024917
[41] Blum K, Ng K C Y, Sato R and Takimoto M 2017 *Phys. Rev. D* **96** 103021
[42] ALICE Collaboration (Acharya S *et al*) 2018 *Phys. Rev. C* **97** 024615
[43] ALICE Collaboration *CERN-LHCC-2013-019*
[44] ALICE Collaboration *CERN-LHCC-2012-012*
[45] ALICE Collaboration *CERN-LHCC-2015-006*
[46] ALICE Collaboration *CERN-LHCC-2013-024*
[47] ALICE Collaboration *CERN-LHCC-2015-021*

IOP Conf. Series: Journal of Physics: Conf. Series **1014** (2018) 012011 doi:10.1088/1742-6596/1014/1/012011

Statistical against dynamical PLF fission as seen by the IMF-IMF correlation functions and comparisons with CoMD model

E.V. Pagano[1], L. Acosta[2,3], L. Auditore[2,4], T. Cap[5], G. Cardella[2], M. Colonna[1], E. De Filippo[2], E. Geraci[2,6], B. Gnoffo[2,6], G. Lanzalone[1,7], C. Maiolino[1], N. Martorana[1,6], A. Pagano[2], M. Papa[2], E. Piasecki[5,8], S. Pirrone[2], G. Politi[2,6], F. Porto[1,6], L. Quattrocchi[2,6], F. Rizzo[1,6], P. Russotto[1], A. Trifiro'[2,4], M. Trimarchi[2,4] and K. Siwek-Wilczynska[9]

[1]INFN Laboratori Nazionali del Sud, Catania, Italy
[2]INFN Sezione di Catania, Catania, Italy
[3]Instituto de Fisica, Universidad Nacional Autonoma de Mexico, Mexico
[4]Dipartimento di Scienze MIFT, Universita' Messina, Italy
[5]National Centre for Nuclear Research, Otwock-Swierk, Poland
[6]Dipartimento di Fisica e Astronomia, Universita' di Catania, Italy
[7]Universita' Kore, Enna, Italy
[8]Heavy Ion Laboratory, University of Warsaw, Poland
[9]Faculty of Physics, University of Warsaw, Poland

E-mail: epagano@lns.infn.it

Abstract. In nuclear reactions at Fermi energies two and multi particles intensity interferometry correlation methods are powerful tools in order to pin down the characteristic time scale of the emission processes. In this paper we summarize an improved application of the fragment- fragment correlation function in the specific physics case of heavy projectile-like (PLF) binary massive splitting in two fragments of intermediate mass(IMF). Results are shown for the reverse kinematics reaction $^{124}Sn+^{64}Ni$ at 35 AMeV that has been investigated by using the forward part of CHIMERA multi-detector. The analysis was performed as a function of the charge asymmetry of the observed couples of IMF. We show a coexistence of dynamical and statistical components as a function of the charge asymmetry. Transport CoMD simulations are compared with the data in order to pin down the timescale of the fragments production and the relevant ingredients of the in medium effective interaction used in the transport calculations.

1. Introduction

Two and multi particles intensity interferometry correlation methods are used to extract space-time information about the emission process [1]. Space-time characteristics of light particles (like protons) emission sources are commonly understood in the frame of the Koonin-Pratt equation [2, 3]. For Intermediate Mass Fragments (IMF) relative momentum correlations, large relative momenta are expected due to the strong influence of the long range Coulomb repulsion in the final decay channel. However, also for IMFs correlation functions, information about the reaction mechanism and time scale could be extracted [4, 5, 6, 7, 8, 9, 10]. The coexistence of dynamical and statistical mechanism in the reaction is an important aspect of the Fermi energy domain where competition between mean field and nucleon-nucleon have to be taken into account in the simulations. First studies using the correlation method were done in the past for central and ternary neck-emissions events with CHIMERA data [11, 12]. In this paper the IMF-IMF correlation function is applied to the physics case as studied by the CHIMERA collaboration, that is the dynamical fission of the projectile-like-fragment (PLF) in 124Sn+64Ni system at 35 A MeV reverse reaction [13, 14]. The basic idea is to provide a time-scale calibration of the IMF-IMF correlation function in that specific case and to study its shape evolution as a function of the mass asymmetry. Notice that in our physics case the detected pairs of particles (IMFs from PLF fission) are associated with large relative momenta as the ones given by the Coulomb repulsions. As a consequence, our data are compared with simulations including Coulomb trajectories calculations. In this work we present comparisons with the CoMD transport model with different parametrizations of the equation of state of nuclear matter (EOS) [15, 16].

2. Experimental analysis

To select projectile-like fragments of fission-like decay, we restrict our analysis to events with IMFs multiplicity equal to two, having both atomic number Z between 3 and 4. Furthermore, we constrained the IMFs parallel velocity to values greater than 5 cm/ns with respect to the beam axis was. Notice that in the studied system the velocity of the center of mass in the laboratory frame is $\approx$ 5.2 cm/ns and the projectile velocity is $\approx$ 8 cm/ns. Complete events were characterized by a total detected parallel momentum greater than 60% of the projectile one and the total collected atomic number Z_{tot} is greater than 40. The charge asymmetry ratio $Z_{Asy} = Z_H \,/Z_L$ of the emitted pairs of IMFs is in the range: $25 \leq Z_H + Z_L \leq 50$, where Z_H and Z_L are the atomic numbers of the heavy and the light IMF of the pair, respectively.The charge asymmetry is adopted in this paper to classify and characterize the exit channel decay properties. So, in summary the selected set of IMFs was decomposed in three partitions: 1 $\leq$

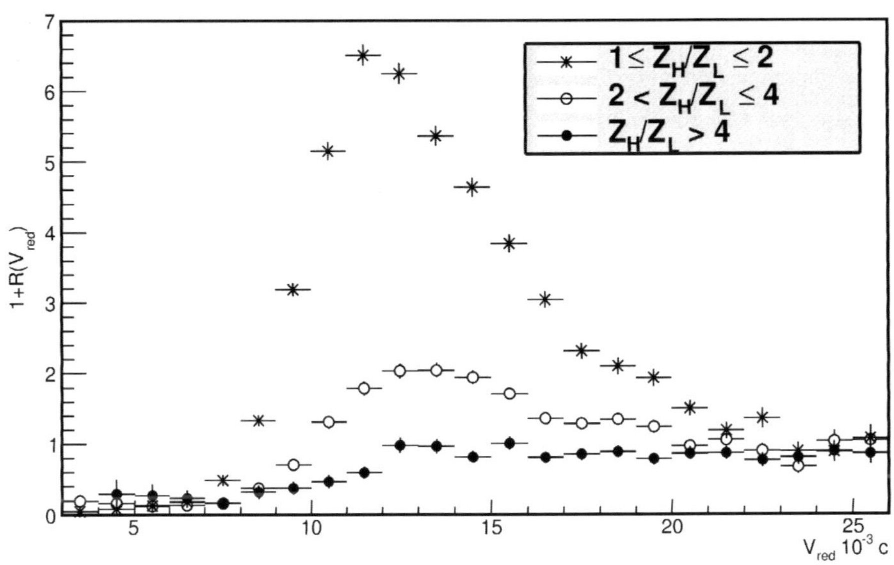

Figure 1. IMF-IMF correlation functions for $25 \leq Z_H + Z_L \leq 50$ gated by different Z_{Asy} condition (see text) [4].

IOP Conf. Series: Journal of Physics: Conf. Series **1014** (2018) 012011 doi:10.1088/1742-6596/1014/1/012011

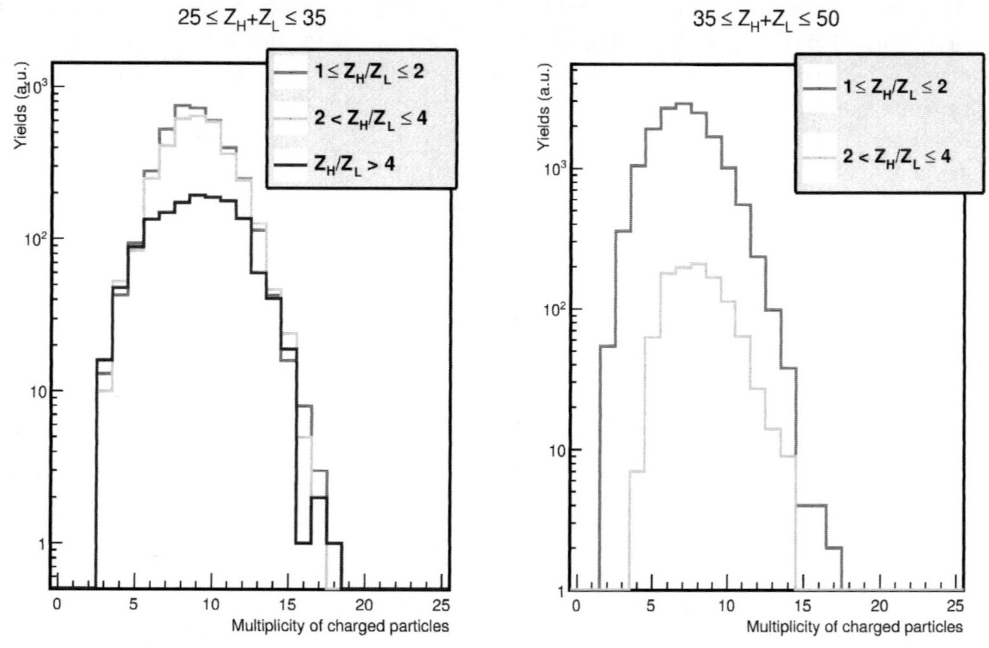

Figure 2. Total charged particles multiplicity as a function of Z_{Asy}, for the two bins of $25 \leq Z_H + Z_L \leq 35$ (left panel) and $35 \leq Z_H + Z_L \leq 50$ (right panel) [4].

$Z_{Asy} \leq 2$, $2 < Z_{Asy} \leq 4$ and $Z_{Asy} > 4$. As it is common, the experimental correlation function

$$1 + R(\mid \mathbf{V_{red}} \mid) = \mathbf{C_{12}} \frac{\mathbf{Y_{Coinc}}(\mid \mathbf{V_{red}} \mid)}{\mathbf{Y_{Uncor}}(\mid \mathbf{V_{red}} \mid)} \tag{1}$$

was evaluated by the ratio between coincident collected pairs of IMFs (in the same event) and uncorrelated ones. The denominator $Y_{Uncor}(\mid \mathbf{V_{red}} \mid)$ was computed with the event mixing technique according to Ref. [17]. In our study event mixing computations have been carefully compared with different techniques such as, pseudo singles computations and Montecarlo pseudo-random so mapping the available phase space configuration. However, our analysis is not based on femtoscopy equations like in Refs [2, 18, 19], but was performed by comparisons with transport simulations and the simulated events have been constrained by the same procedures applied to the experimental data. The three IMF-IMF correlation functions, as a function of the three ranges of theZ_{Asy}, are shown in Fig.1 and they have been evaluated as a function of the reduced velocity defined as it follows [6]:

$$\mathbf{V_{red}} = \frac{\mathbf{V_H} - \mathbf{V_L}}{\sqrt{\mathbf{Z_H} + \mathbf{Z_L}}} = \frac{\mathbf{V_{rel}}}{\sqrt{\mathbf{Z_H} + \mathbf{Z_L}}} \tag{2}$$

This quantity allows to add together pairs of particles having different atomic numbers [6] [6] in order to increase the statistics that plays a crucial role in such studies. The IMF-IMF correlation functions, for the three ranges of charge asymmetry show very different shapes. Notice that the well-defined bump observed for $1 \leq Z_{Asy} \leq 2$ is centered at a value of the reduced relative velocity that is in agreement with a binary sequential PLF splitting, as it is deduced by the Coulomb repulsion in agreement with the Viola systematics [20]. The energy dissipation has been analysed by inspecting the charged particle multiplicity in the three ranges

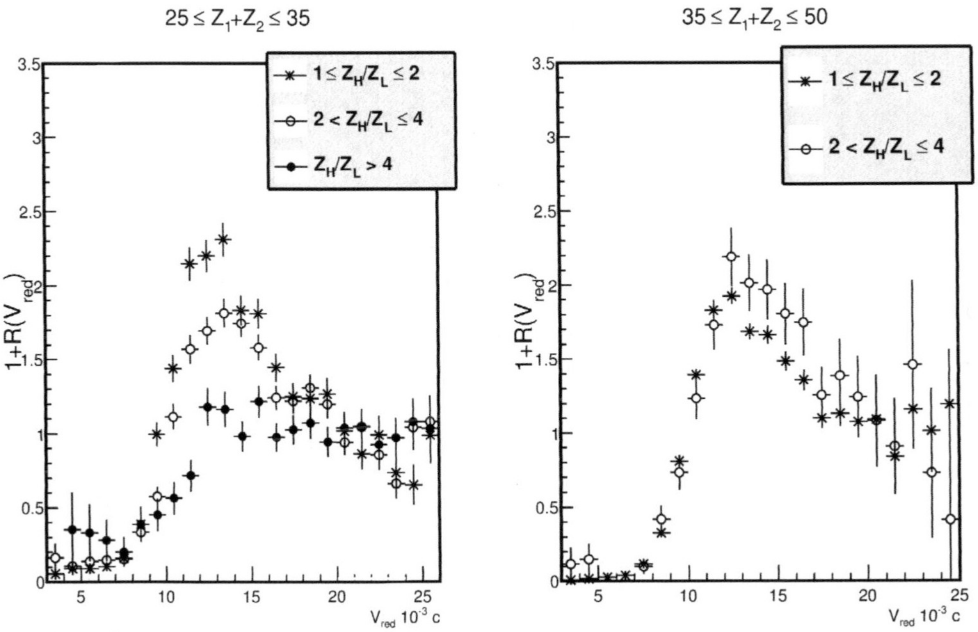

Figure 3. IMF-IMF correlation function for different Z_{Asy} : $25 \leq Z_H + Z_L \leq 35$ (left panel) and $35 \leq Z_H + Z_L \leq$ (right panel) [4].

of charge asymmetry and the corresponding sum $Z_H + Z_L$. We evidenced the tendency of the dissipation to grow up with the increase of the asymmetry from a most probable value around 7 charged particles, for the most symmetric gate, up to about 10 charged particles for the most asymmetric one.

In order to achieve a more careful understanding of the different observed shapes (and assuming that the sum $Z_H + Z_L$ is a measure of the size of the emitting source) we improve the class of the selected events with respect to the size of the emitting system and (consequently) the range of the energy dissipation. We have considered two complementary narrower bins of the sum of the atomic number of the two pairs , i.e., one bin defined in the range, $25 \leq Z_H + Z_L \leq 35$ (left panel of Fig.2) and the other one in the range, $35 \leq Z_H + Z_L \leq 50$ (rightpanel of Fig.2).

In Fig.3 the new results for the IMF-IMF correlation functions are shown: in the left panel for the most dissipative range: $25 \leq Z_H + Z_L \leq 35$ and in the right panel for the less dissipative one: $35 \leq Z_H + Z_L \leq 50$. In the first case (larger dissipation) hierarchies in both intensity and shape as a function of the Z_{Asy} are preserved; in contrast, in the second case (lower dissipation) the correlation functions show no significant differences as a function of the Z_{Asy}, (the value of the asymmetry $Z_{Asy} > 4$ is not shown because the statisticswas not significant). The natural interpretation is that in the most dissipative range the changing shape of the correlation function is a clear indication of the evolutionary time scale character of the nuclear mechanism leading to the production of the IMFs. Instead, for the most gentle collisions associated with sizes in the range $35 \leq Z_H + Z_L \leq 50$ the space-time configuration is only poorly affected by the asymmetry parameter (practically, both shape and intensity of the correlation functions are unchanged). Once the size of the emitting source was fixed by the value of $25 \leq Z_H + Z_L \leq 35$, the evolution of the intensity and shape of the correlation function vs. the charge asymmetry could be linked to the time scale of the decay process from the prompt dynamical decay (fast process) up to

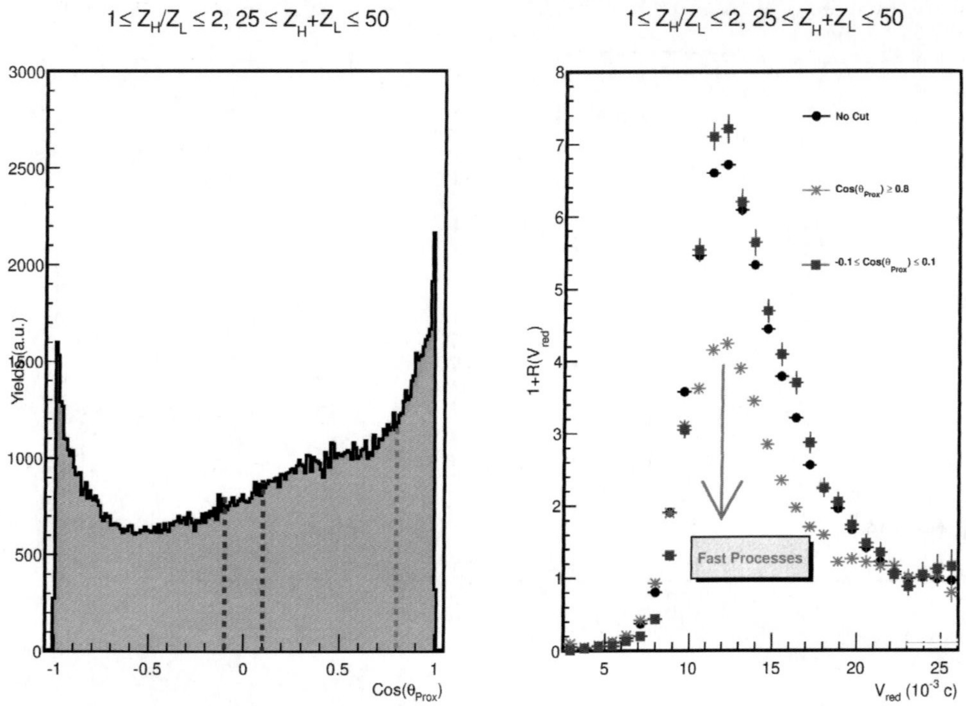

Figure 4. (left panel) distribution of the cosines of θ_{prox} for $25 \leq Z_H + Z_L \leq 50$ and $1 \leq Z_{Asy} \leq 2$. (Right panel) IMF correlation functions with and without constrains in $cos(\theta_{prox})$.

the equilibrated fission decay (sequential process). The correlation functions were also studied as a function of the theta proximity angle θ_{prox}, which is an independent indicator of both equilibrium emission and dynamical processes [21].

In brief, the theta proximity represents the angular distribution of a selected light IMF of the pair with respect to the reconstructed center of mass vector velocity of the emitting primary PLF source. In the left panel of the Fig.4 the distribution of the $cos(\theta_{prox})$ for the full range $25 \leq Z_H + Z_L \leq 50$ and for the most symmetric gate of charge asymmetry is shown. In a pure sequential fission decay such a distribution should be forward-backward symmetric with respect to the value of $cos(\theta_{prox})$=0. So, in agreement with the analysis of refs. [13, 14] the distribution of Fig.4 (left panel) is dominated by equilibrated sequential PLF fission decay and only a small but sizeable contribution ($\approx 10\%$) of it is associated with a prompt dynamical process evidenced by a forward angular asymmetry for the events with a $cos(\theta_{prox})$ close to 1.

It is very interesting to test the sensitivity the correlation function to the $cos(\theta_{prox})$ asymmetry. In the right panel of Fig.4 three correlation functions have been evaluated with and without constraints in $cos(\theta_{prox})$ distribution. All the three correlations are peaked at nearly the same value of the reduced relative velocity (see Fig.1 and text). The correlation function with the largest correlated intensity (blue-full square) corresponds to the one obtained constraining the $cos(\theta_{prox})$ (bin of 0.2 width) in the range of values -0.1 $\leq cos(\theta_{prox}) \leq$ 0.1 where a pure statistical fission-like sequential decay was evidenced [13, 14] and it is identical (within the accuracy) to the one (black-full dots) corresponding to the total $cos(\theta_{prox})$ distribution. The third correlation function (red-stars) is calculated constraining the total $cos(\theta_{prox})$ distribution for values (bin of 0.2 in width as in the black full dots one) $cos(\theta_{prox}) \geq$ 0.8, in the region where the strong forward-backward asymmetry shows the largest contribution of a dynamical-out of equilibrium fission-like process. It is clearly noticed that the intensity of this latter correlation function is strongly decreased and its shape is broadened; this is a further strong and complementary indication of the high sensitivity of the correlation function to the prompt/dynamical processes.

3. Transport simulations and comparisons

To complete our analysis, we made comparisons between the experimental data and the CoMD model calculations [15, 16]. In Fig.5, e.g., for charge asymmetry in the range, $1 \leq Z_{Asy} \leq 2$, detailed comparisons between the experimental and simulated events of the main experimental observables (a): $Z_H + Z_L$; b): Z_{Asy} and c): $cos(\theta_{prox})$ distributions)used in this analysis are shown for different parametrization of EOS.

Notice that in CoMDII [15], as already used in the preliminary analysis of refs [4, 5], the sequential decays of the primary fragments have been also included. In CoMDIII [16] simulations and comparisons with experimental data shown in Fig.5 (a), b) and c)) the sequential decay is not included in the simulations. The simulated events have been constrained as the experimental data and, remarkably, the range of the resulting simulated impact parameters, i.e., about $2 \leq$ b ≤ 8 fm, was found to be consistent with the evaluated experimental one, as deduced by using the Cavata method [22].

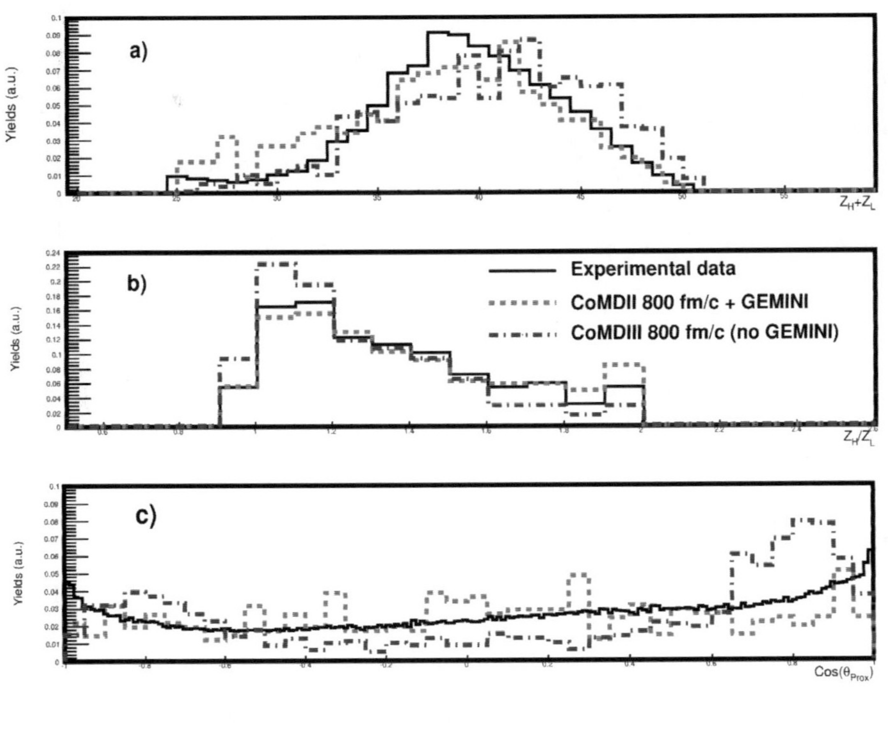

Figure 5. Comparisons between simulated and experimental observables (see text).

In Fig.6 the comparison between the IMF-IMF experimental correlation functions (in the figure the three ranges of Z_{Asy}, like in Fig.1, are reported for simplicity) and the simulated ones for the largest range of the charge sum, $25 \leq Z_H + Z_L \leq 50$ is shown. Full squares (in red) represents a time computation of 800 fm/c in CoMDIII simulations and the full dots (in green) time computation of 650 fm/c. A good agreement between data and simulations is observed for the intermediate and largest asymmetries, indicating that, the case of dynamical breakup/fission occurring within 800 fm/c is dominated by an asymmetric splitting of intermediate or large charge asymmetries. It is also envisaged to continue the comparison using the CoMD and also with other transport models in order to obtain information on the reaction mechanism and in medium effective interaction that are as much as possible model independent. For instance, a good candidate to simulate dynamical fission processes is the stochastic mean field models, such as SMF and BLOB [23, 24].

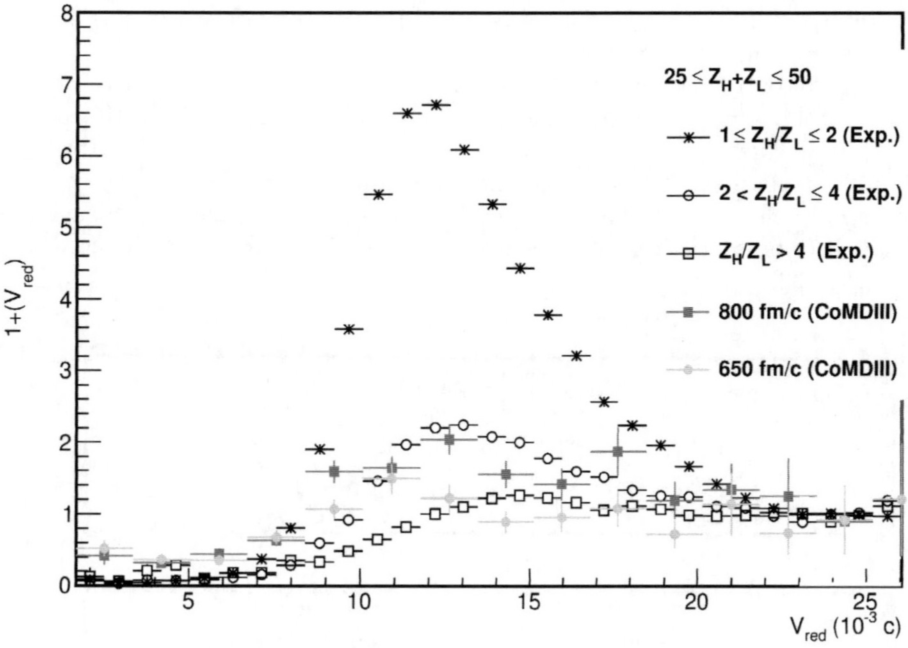

Figure 6. Comparisons between theoretical and experimental correlation functions.

4. Conclusions and future perspectives

In this work a first detailed study of the IMF-IMF correlation functions of particle pairs of both high relative momenta and charges, in the specific case of dynamical PLF fission in the collision $^{124}Sn + ^{64}Ni$ at 35 AMeV [14] [13] has been presented. We have shown that the correlation function for massive fragments is a powerful and complementary experimental method in order to pin down on the time scale, size and dissipative mechanism of multiple massive IMF emission in nuclear reactions at Fermi energies. Important sources of information are expected to be available by careful comparisons of experimental data with advanced nuclear dynamics simulations. The results discussed in this paper strongly encourage for further detailed investigations with new exclusive experiments in this field [25].

[1] Henzl V *et al.* 2012 *Phys. Rev. C* **85** 014606
[2] Koonin S E 1977 *Phys. Lett. B* **70** 4
[3] Pagano E V 2013 *Nuovo Cimento C* **36 (issue 4)** 9
[4] Pagano E V *et al.* 2016 *Il Nuovo Cimento C* **39** 404
[5] Pagano E V A.A. 2016/2017 *Phd Thesis Univerity of Catania*

[6] Kim Y D *et al.* 1992 *Phys. Rev. C* **45** 338
[7] Schapiro O, DeAngelis A R and Gross D H E 1994 *Nucl. Phys. A* **568** 333
[8] Pal S 1995 *Nucl. Phys. A* **594** 156
[9] Xiao Z G *et al.* 2006 *Phys. Lett. B* **639** 436-440
[10] Pagano E V *et al.* 2017 *PoS(BORMIO)* 022
[11] Geraci E *et al.* 2004 *Nucl. Phys. A* **732** 173
[12] Maiolino C *et al.* 2005 *Proceedings of the XLIII International Winter Meeting on Nuclear Physics, Bormio, edited by Iori I and Bortolotti A* **124** 194
[13] De Filippo E *et al.* 2005 *Phys. Rev. C* **71** 064604
[14] Russotto P *et al.* 2010 *Phys. Rev. C* **81** 064605
[15] Papa M, Maruyama T and Bonasera A 2001 *Phys. Rev. C* **64** 024612
[16] Papa M 2013 *Phys. Rev. C* **87** 014001
[17] Lisa M A, Gong W G, Gelbke C K and Lynch W G 1991 *Phys. Rev. C* **44** 2865
[18] Lednicky R and Lyuboshitz V L 1982 *Sov. J. Nucl. Phys.* **35** 770
[19] Pratt S and Ztang B 1987 *Phys. Rev. C* **6**
[20] Viola V E, Kwiatkowski K and Walker M 1985 *Phys. Rev. C* **31** 1550
[21] De Filippo E *et al.* 2012 *Phys. Rev. C* **86** 014610 and references therein.
[22] De Filippo E and Pagano A 2014 *Eur. Phys. J. A* **50** 32
[23] Rizzo C *et al.* 2014 *Phys. Rev. C* **90** 054618
[24] Napolitani P, Colonna M 2016 *EPJ Web of Conferences* **117** 07008
[25] Defilippo E, Pagano E V, Russotto P *et al.* CHIFAR @LNS, approved proposal.

IOP Conf. Series: Journal of Physics: Conf. Series **1014** (2018) 012012 doi:10.1088/1742-6596/1014/1/012012

The RIB facility EXOTIC and its experimental program at INFN-LNL

Concetta Parascandolo
INFN - Sezione di Napoli, Via Cintia, I-80126 Napoli, Italy
E-mail: concetta.parascandolo@na.infn.it

Abstract. In this contribution, I will present a review about the EXOTIC facility and the research field accessible by using its Radioactive Ion Beams. The EXOTIC facility, installed at the INFN-Laboratori Nazionali di Legnaro, is devoted to the in-flight production of light Radioactive Ion Beams in the energy range between 3-5 MeV/nucleon. The scientific activity performed at EXOTIC concerns different aspects of nuclear physics and nuclear astrophysics, such as, the investigation of reaction mechanisms and nuclear structure, resonant scattering experiments and measurements of nuclear reaction cross sections of astrophysical interest.

1. Introduction

During the last decades, the development of Radioactive Ion Beam (RIB) Facilities opened the gates to the investigation of fundamental studies in physics related to nuclei far from the valley of β-stability. These nuclei might exhibit exotic features, such as, for instance: (i) a very low binding energy (typically 1.0 MeV), (ii) the last neutrons mostly orbiting in the outer part of the nucleus in a "skin-like structure", (iii) a "halo" structure corresponding to a rather inert core plus one or two neutrons (protons) circulating around in an orbit whose radius is much larger than that observed for stable well-bound nuclei close to the valley of the β-stability.

These peculiar charateristics give rise to a rather wide variety of phenomena, such as new radioactive decay modes, e.g. one-proton, two-proton and neutron radioactivity, and also the shell closures show an evolution going from the valley of the β-stability to the neutron- and proton-drip lines. Moreover, experiments with RIBs allow to explore the properties of isotopes with a proton-to-neutron ratio very different from the stable ones, measure reaction cross sections of astrophysical interest occurring in explosive environments, constrain the isospin-dependent nucleon-nucleon interaction in neutron-rich nuclei and in neutron stars, synthesize superheavy elements and test physics beyond the standard model.

2. EXOTIC facility and Experimental Set-up

The EXOTIC facility [1, 2, 3, 4] has been commissioned at INFN-Laboratori Nazionali di Legnaro (INFN-LNL) in the years 2003-2004 and upgraded in 2012 [5]. This facility is dedicated to the in-flight production of low-energy light RIBs, by inverse kinematics nuclear reactions induced by the intense heavy-ion beams delivered from the INFN-LNL Tandem XTU accelerator hitting a light gas target, such as H_2, D_2, ^{3}He and ^{4}He.

The main features of the facility, shown in Fig. 1, are a large RIB acceptance of the optics elements and a large capability to suppress all the undesired scattered beams. At the beginning, there is a production gas target that consists of a 5 cm-long cylindrical cell walled with windows made of 2.2 μm thick Havar foil, operating at room or at liquid N_2 temperatures at pressures up to 1 atm. Then, after the production, the beam can be selected and transported through a system made of: a triplet of large diameter quadrupole lenses, a 30° bending magnet, a Wien filter and a second triplet of quadrupole lenses, placed before the tracking detectors and the reaction chamber. The Wien filter eliminates to a very large extent the tails of the primary beam that pass through the system with the same magnetic rigidity. In order to stop the RIB contaminants, different slit sets are installed along the beamline and can be adjusted to the envelope of the produced RIB. The facility has a length of 8.34 m and presents the following charateristics: $\Delta E/E = \pm 10\%$, $\Delta p/p = \pm 5\%$, $\Delta\theta = \pm 50$ mrad, $\Delta\phi = \pm 65$ mrad, $\Delta\Omega \approx 10$ msr, $B\rho = 0.98$ Tm.

So far, RIBs of ^{7}Be, ^{8}B, ^{17}F, ^{15}O, ^{8}Li, ^{10}C and ^{11}C at 3-5 MeV/nucleon have been delivered with intensities about 10^6, 10^3, 10^5, $4\text{x}10^4$, 10^5, $5\text{x}10^3$ and $2\text{x}10^5$ pps, respectively, and with 98-99% purity (except for ^{8}B that has a lower purity).

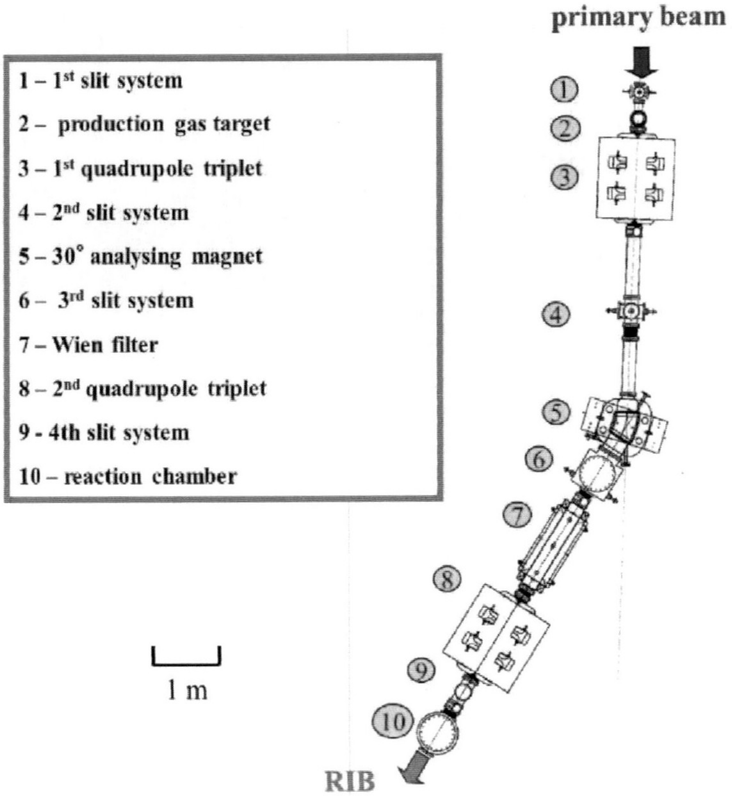

Figure 1. Layout of the EXOTIC facility.

The experimental set-up [6, 7] of the EXOTIC facility is presented in Fig. 2. It is situated at the end of the beamline and consists of: a) the RIB tracking system and b) the detection system, EXPADES, a new charged-particle telescope array. EXPADES is a portable array and is flexible to suit many experiments. It can be also coupled with γ-ray and neutron arrays.

The Parallel Plate Avalanche Counters (PPACs) of the tracking system are position-sensitive and fast detectors, which can sustain counting rates up to $\sim 10^6$ Hz. They are placed 909 mm

IOP Conf. Series: Journal of Physics: Conf. Series **1014** (2018) 012012 doi:10.1088/1742-6596/1014/1/012012

(PPAC A) and 365 mm (PPAC B) upstream the reaction target, at the EXOTIC focal plane. The PPACs are filled with C_4H_{10} gas at a working pressure of 10-20 mbar. Each PPAC has an active area of 62x62 mm^2 and is sealed with 1.5 μm-thick mylar windows. The detector has a central cathode and two anodes, placed symmetrically at a distance of 2.4 mm from the cathode. The cathode is a unique plate of 1.5 μm-thick stretched mylar foil, while each anode is a mesh of 60 gold-plated tungsten 20 μm-thick wires in the x and y directions, with a spacing of 1 mm. The position information of a particle crossing the PPAC is extracted from the anode signals by using a delay-line readout. The 1 mm resolution of the two PPACs allows us to reconstruct the position of the event on the reaction target with a 2.3 mm position resolution. The cathode signal is used as a reference time for Time of Flight measurements (giving a START signal with 0.9 ns FWHM time resolution) and for trigger purposes.

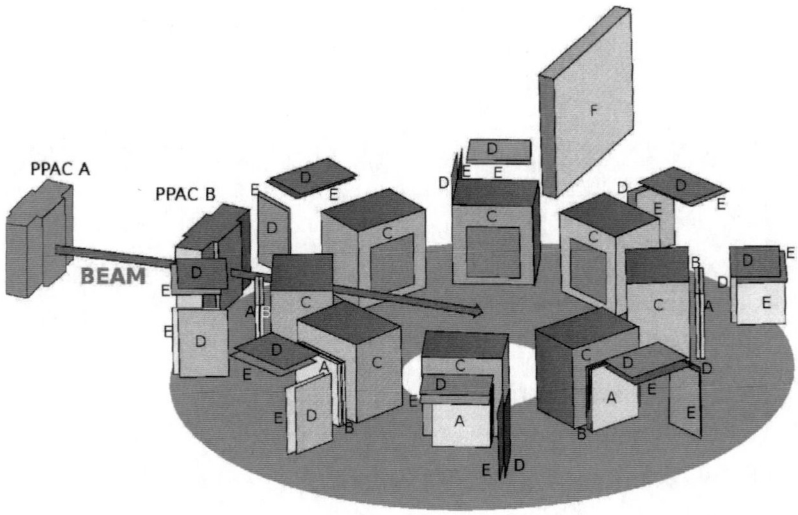

Figure 2. The position sensitive PPACs for the RIB tracking system and the detection system, EXPADES, placed in the reaction chamber (element 10 of Fig. 1) at the end of the EXOTIC facility.

EXPADES is an array of eight telescopes arranged in a cylindrical configuration around the reaction target (Fig. 2). The telescope structure is flexible and is composed of two Double Side Silicon Strip Detectors (DSSSDs) and/or Ionization Chamber (IC), depending on the experimental requests. We use 40/60 μm-thick DSSSDs for the ΔE stage (B in Fig.2) and 300 μm-thick DSSSDs for the E_{res} one (A in Fig. 2). Each DSSSD has 64-mm long 32x32 strips, with 2 mm pitch size and 40 μm interstrip separation, defining thus a $\sim$ 2x2 mm^2 pixel structure. For experiments requiring the detection of more energetic particles than those stopped in the E_{res}, 1 mm-thick DSSSDs can substitute the 300 μm-thick ones or can be used in addition to the previous stages. ASIC-based electronics was employed for the treatment of the E_{res} signals, with a high granularity and a very low cost. For the signal readout of the ΔE stage a compact low-noise electronics with very good energy and timing characteristics was developed by our collaboration. A valid alternative to allow the ΔE-E_{res} identification of reaction products with range in silicon shorter than 40/60 μm is the use of a home-made transverse field IC (C in Fig. 2) that can give the possibility to tune the effective thickness of the ΔE layer by changing the gas pressure. The IC has 65x65 mm^2 entrance and exit windows (1.5 μm-thick mylar) and can be filled with CF_4 gas at pressures up to 100 mbar. The low-noise charge-sensitive preamplifiers for the ΔE DSSSDs (D in Fig. 2), those of the ICs (not present in Fig. 2) as well as the electronic boards (E in Fig. 2) for the E_{res} DSSSDs are placed under vacuum close to the detectors. In this way, we manage to keep as low as possible the DSSSDs electronic thresholds, typically 300-500 keV.

Fig. 3 shows one of the possible configurations of the detection system installed in the reaction chamber of the EXOTIC facility: in this case, six two-stage DSSSD telescopes of EXPADES array are used.

Figure 3. Six two-stage DSSSD telescopes of EXPADES array mounted in the reaction chamber of the EXOTIC facility. The RIB comes from the left side, passing through PPAC A and PPAC B (shown in the picture) of the tracking system.

3. Experimental program and Perspectives

The experimental program at EXOTIC is varied and aims at different topics of nuclear physics and nuclear astrophysics.

3.1. Reaction Mechanisms

The scientific activity at EXOTIC facility started with the study of reaction mechanisms induced by light radioactive (exotic) nuclei impinging on medium- and heavy-mass targets at energies near to the Coulomb barrier.

The peculiar features of these nuclei influence the reaction mechanisms giving a picture that is rather different from that of well bound species because of the appearance of new phenomena. The loose binding of the weakly bound nuclei yields in a strong breakup and/or transfer channel which could influence the elastic scattering (i.e. the total reaction cross section) and the fusion cross section. Indeed, different theoretical moldels have predicted both enhancement and hindrance of the sub-barrier fusion cross section due to this coupling. Several review articles have recently been written on this topic [20, 21, 22, 23, 24, 25]. From an experimental point of view, many studies with weakly bound stable and radioactive species (RIBs) were undertaken to disentangle the different reaction mechanisms, also at the EXOTIC facility [8, 9, 10, 11, 12, 13, 14, 15, 16, 17, 18, 19].

In 2010 we started the study of ^{7}Be-induced reactions. This RIB can be produced at EXOTIC facility through the charge exchange reaction ^{1}H(^{7}Li, ^{7}Be)n and, then, selected by a careful tuning of the quadrupole lenses, the dipole magnet and the Wien filter of the facility. This projectile was chosen because is a weakly-bound proton-rich nucleus with a very pronounced ^{3}He - ^{4}He cluster configuration in its ground state. The ^{7}Be (S_{α} =1.586 MeV) represents, among all light nuclei, the cleanest case where the study of the breakup/transfer interplay into the reaction dynamics can be addressed in detail. Indeed, the two ^{7}Be inner clusters are very stable fragments with similar masses, while all breakup studies previously performed always involved the complicated and low-efficiency detection of neutrons (like in the case of the ^{6}He and ^{9,11}Be breakup processes), the emission of a weakly-bound fragment (e.g., the deuteron emission in the ^{6}Li breakup), the emission of a radioactive fragment (as for the triton emission in the ^{7}Li breakup) or a very small breakup probability, as in the case of the ^{17}F breakup process into ^{16}O + p. Moreover, the ^{7}Be nuclear structure is quite similar to that of its mirror nucleus ^{7}Li (S_α = 2.468 MeV), while its binding energy is rather close to that of the lightest particle stable lithium isotope ^{6}Li (S_α = 1.475 MeV). A better comparison between the ^{7}Be and ^{6,7}Li reaction dynamics at Coulomb barrier can easily be achieved thanks to the large amount of data available for the reactions induced by the stable projectiles on several different targets.

First, the energy and angular distributions of ^{3}He and ^{4}He ions produced in the ^{7}Be + ^{58}Ni reaction at a bombarding energy of 22 MeV have been measured for the first time. The quasi-elastic data, displayed in Fig. 4, were in remarkable agreement with an earlier measurement performed at the University of Notre Dame [26], while the ^{4}He production (measured for the first time) turned out to be 4-5 times more abundant than that of its lighter counterpart. This result clearly rules out the possibility that in this energy range the ^{7}Be reaction dynamics is dominated by the exclusive breakup process ^{7}Be → ^{3}He + ^{4}He (S_α= 1.586 MeV), otherwise similar yields for the two helium isotopes would have been observed. Unfortunately, due to small geometrical efficiency and limited beam-time availability, no coincidences between projectile fragments were recorded. Therefore, extensive theoretical calculations were undertaken in order to determine the possible contributions of the n-stripping, n-pickup and exclusive breakup processes to the inclusive production cross section of ^{3}He and ^{4}He [10].

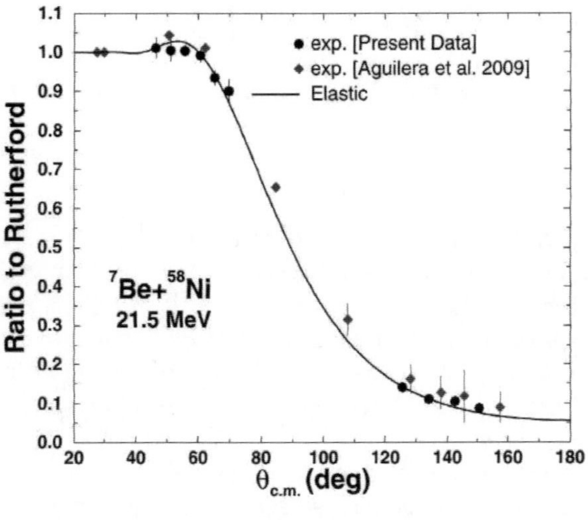

Figure 4. Quasi-elastic angular distribution for the ^{7}Be + ^{58}Ni system at 21.5 MeV. Black circles represent the present measurement, while blue diamonds are taken from an earlier measurement performed in [26]. The solid black line denotes the result of an optical model calculation, as explained in [10].

Recently, we performed the ^{7}Be + ^{208}Pb reaction for the first time at three near-barrier energies. The goals of this experiment were to measure for the first time the near-barrier quasi-elastic angular distribution and to detect (at least) a few coincidences between projectile

IOP Conf. Series: Journal of Physics: Conf. Series **1014** (2018) 012012 doi:10.1088/1742-6596/1014/1/012012

fragments, in order to get a deeper insight on the reaction dynamics induced by this exotic projectile. The experiment benefited from the upgrade of our facility EXOTIC [5], which is now able to deliver ^{7}Be beams energetic enough to match the energy range around the Coulomb barrier also for reactions on heavy targets. We measured the quasi-elastic scattering differential cross section (shown in Fig. 5) and extracted the total reaction cross section. The comparison with the reaction cross sections obtained for the mirror projectile ^{7}Li and the similarly weakly-bound nucleus ^{6}Li interacting with the same target clearly suggests that the ^{7}Be data essentially follow the trend individuated by the heavier lithium isotopes. This result indicates that, for these projectiles, nuclear structure might play a more relevant role in the reaction dynamics rather than the projectile binding energy. In addition, we started the investigation of the energy spectra and angular distributions for the inclusive production of ^{3}He and ^{4}He nuclei. We immediately observed that the ^{4}He production is a factor 4-5 higher than the ^{3}He one. This confirms that also for the interaction with heavy targets the ^{7}Be reaction dynamics at near barrier is not dominated by the breakup process, as found in the ^{7}Be + ^{58}Ni system. The detailed analysis of these events is presently in progress.

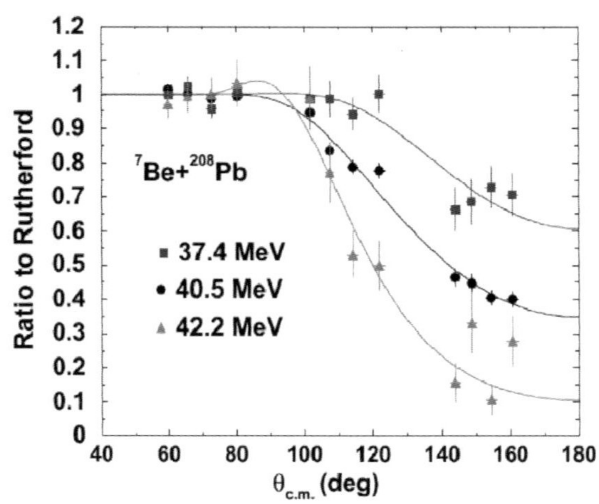

Figure 5. Quasi-elastic angular distribution for the ^{7}Be + ^{208}Pb system at three beam energies: 37.4 MeV (squares), 40.5 MeV (circles) and 42.2 MeV (triangles). Lines are the results of a best-fit analysis of the experimental data within the framework of the optical model, as described in detail in [12].

3.2. Clustering

Clustering phenomena [27] are well known in nuclear physics for stable nuclei, both α-conjugate (N=Z, A=2N), like ^{8}Be, ^{16}O, ^{20}Ne, and non-α-conjugate, like ^{6}Li and ^{7}Li. In general, it is expected that light exotic nuclei may also exhibit cluster behavior. Moving out of the valley of stability, configurations can be found where at least one of the clusters is unbound or weakly bound, thus not satisfying the strong internal correlation requirement of classical clusters. This is the so-called *exotic clustering* regime. The study of such systems presents many difficulties, due to, mainly, the low intensities typical of RIBs. Therefore, few significant experimental studies have been performed so far.

We started to study α-clustering phenomena in light exotic nuclei, employing the Thick Target Inverse Kinematics (TTIK) scattering technique [28], with the RIB impinging on a ^{4}He gas target. As the projectiles slow down in the gas, elastic scattering reactions across a wide

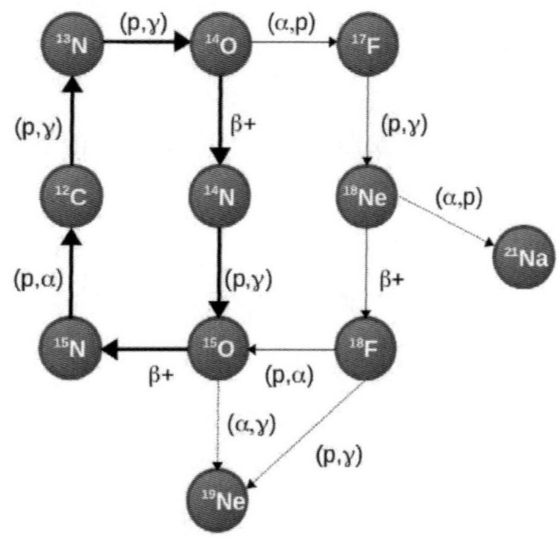

Figure 6. The hot-CNO cycles. (Bold arrows) Dominant reaction path; (light arrows) at T = 0.4 GK the sequence started by $^{14}O(\alpha,p)^{17}F$. At still higher temperatures, it may be possible to leave the cycle (dashed arrows).

range of center-of-mass energies can take place (corresponding to a range of depths within the gas). The pressure of the gas is tuned such that the RIB completely stops in the gas, allowing a detector to be placed at 0° , i.e., on the beam axis, while the energetic recoiling light target nuclei, due to their low rate of energy loss, can traverse the gas and be detected. The TTIK method is useful for measurements with low-intensity RIBs since it allows to measure the elastic scattering excitation function and the reduced α-width directly, over a wide energy range by using a single beam energy, since the gas acts as a target but also as a degrader. In the following case, the technique has been refined by using Time of Flight information [29, 30] to remove contributions from processes other than elastic scattering [31]. Moreover, modifications of the EXOTIC facility were performed in early 2015, to allow the realization of experiments by employing RIBs impinging on reaction gas targets. Indeed, a new small chamber was built hosting the PPAC B that separates, through a 2.2 μm thick Havar window, the scattering chamber (filled with ^{4}He gas) from the beamline (at high vacuum).

In particular, we searched for α-cluster states in ^{19}Ne above its α-decay threshold measuring, for the first time, the $^{15}O(^4He, ^4He)^{15}O$ elastic scattering excitation function with the TTIK technique. The 28.5 MeV ^{15}O RIB was produced with the EXOTIC facility by means of the in-flight technique via the $p(^{15}N,^{15}O)n$ reaction, with the ^{15}N beam delivered by the INFN-LNL Tandem XTU accelerator.

^{19}Ne has attracted much interest due to the astrophysical importance of the $^{15}O(\alpha,\gamma)^{19}Ne$ reaction, identified as a break-out from the hot-CNO cycles (shown in Fig. 6) and a trigger for the following *rp*-capture populating masses higher than neon [32]. Neon isotopes exhibit strong α-clustering phenomena, even in the ground states. Therefore, the idea has been to investigate the possible cluster nature of the neutron-poor side of the neon isotopes, almost totally unknown [33]. After the elastically scattered α particles have been selected, the final excitation function is obtained and shown in Fig. 7. The excitation function has been analysed by using a full R-Matrix fit performed using Azure2 code [34], revealing a significant number of new resonances. The results show evidence for ^{15}O + α cluster structure which could explain the increased role played by some levels in the astrophysical ^{15}O(α,γ)^{19}Ne reaction in the break-out from the hot-CNO cycles.

We extended this research field to the neutron-poor side of the oxygen isotopes, that is, at the moment, almost totally unknown concerning the clustering point of view, while neutron-rich oxygen isotopes have been intensively investigated for their cluster nature. By using the same technique as before, recently we investigated the presence of possible α-cluster states in ^{15}O measuring the elastic scattering excitation function of the system ^{11}C + ^{4}He with a ^{11}C RIB produced by the EXOTIC facility. The analysis is still ongoing.

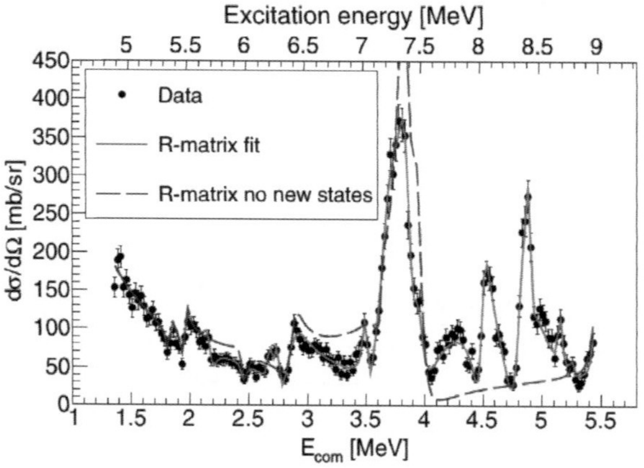

Figure 7. Excitation function for ^{19}Ne at 0°. The data points are overlaid by the R-matrix fit (solid red line). The fit without the newly observed levels is shown for comparison (dashed blue line). See [33] for more details.

3.3. Nuclear Astrophysics

Another possibility offered by the EXOTIC facility is to perform measurements of astrophysical interest with RIBs impinging on solid or gas light targets in inverse kinematics. Among the different processes of stellar nucleosynthesis forming elements heavier than ^{9}Be, the *rp*-capture and αp processes, occurring in explosive astrophysical environments, are those than can be investigated by using the EXOTIC RIBs.

We want to develop a radioactive ^{18}Ne beam through the ^{3}He(^{16}O,^{18}Ne)n reaction. The ^{18}Ne RIB is of primary importance for nucleosynthesis of heavier elements. Indeed, it allows the break-out from the second hot-CNO cycle through the ^{18}Ne(α,p)^{21}Na reaction (see Fig. 6), converting the initial CNO isotopes into heavier elements [32]. However, the actual astrophysical conditions under which this occurs depend critically on the ^{18}Ne(α,p)^{21}Na rate. This rate has been studied through direct and indirect measurements (see [35] and references therein).

The importance for nuclear astrophysics of the ^{18}Ne(α,p)^{21}Na reaction and the disagreement between direct and indirect measurements call for further investigation. For these reasons, it is important to perform a new measurement in the overlapping region, where the different experiments disagree, and to extend toward lower energies, thus populating the region of interest for hot-CNO breakout in x-ray bursts. We plan to perform a direct measurement of the cross section for the ^{18}Ne(α,p)^{21}Na reaction with an extended gas target at the EXOTIC facility in the next future, after testing the ^{18}Ne production at our facility.

Another measurement relevant to the astrophysics can be performed at EXOTIC such as the ^{30}P(p,γ)^{31}S with a ^{30}P RIB. The ^{30}P(p,γ)^{31}S reaction is thought to be the bottleneck for the production of heavy elements (from Si to Ca) in the explosion of ONe novae due, in part, to the long β decay half life ($t_{1/2}$= 2.5 min) of ^{30}P, which is comparable to the duration of nova nucleosynthesis (Fig. 8). Consequently, uncertainties in the thermonuclear ^{30}P(p,γ)^{31}S reaction rate affect the interpretation of several observables that could elucidate the astrophysics of novae [36]. A related problem involves the anomalously high ^{30}Si/^{28}Si rate in pre-solar grains of possible ONe novae origin [37]. Uncertainties in the ^{30}P(p,γ)^{31}S reaction rate hinder accurate quantitative comparisons to the silicon isotopic ratios from nova models.

Up to now, there is enough experimental information about the ^{30}P(p,γ)^{31}S reaction thanks to various indirect measurements (see [38] and references therein). Unfortunately, comparing various interpretations of this information reveals ambiguities and discrepancies. Direct measurements with ^{30}P RIB and hydrogen targets are not currently possible due to the lack of sufficiently intense, high-quality ^{30}P beams at the relevant energies. Therefore, in future, we intend to develop this RIB at EXOTIC, through the ^{4}He(^{27}Al, ^{30}P)n or ^{2}H(^{29}Si,^{30}P)n reactions, in order to perform a direct measurement and try to give relevant information to this astrophysical quest.

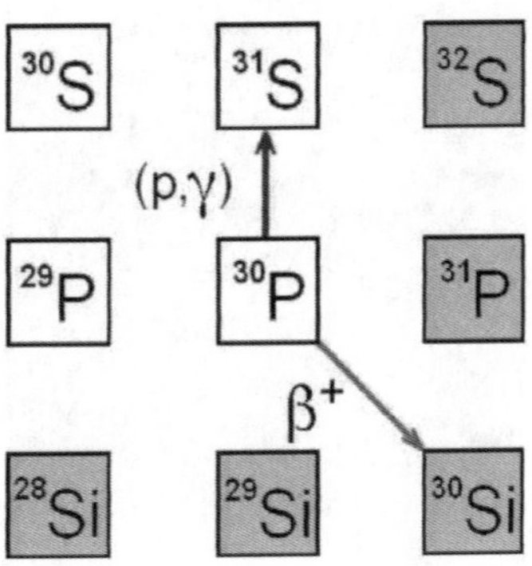

Figure 8. Portion of the chart of nuclides showing the competition between the ^{30}P(p,γ)^{31}S reaction and the ^{30}P β decay. The abscissa is the neutron number N and the ordinate is the proton number Z. Stable nuclides are represented in grey and unstable nuclides are represented in white [38].

Moreover, experiments based on the Trojan Horse Method (THM) [39, 40, 41, 42] are considered. In this direction, a first experiment was performed to investigate the ^{7}Be(n,α)^{4}He reaction to shed light on the cosmological lithium-problem, of interest in the Big Bang Nucleosynthesis [43]. To this aim, the THM has been applied to the quasifree reaction ^{2}H(^{7}Be,α ^{4}He)p, by using a 20.4 MeV ^{7}Be RIB impinging on a CD_2 target. By referring to the polar diagram of Fig. 9, deuteron represents the TH-nucleus, undergoing its breakup in neutron (participant) and proton (spectator). In the same research program, an independent experiment has been performed at Center-for-Nuclear-Study Radioactive Isotope Beam (CRIB) (Riken, Japan) [44, 45] to study simultaneously the ^{7}Be(n,p)^{7}Li and ^{7}Be(n,α)^{4}He reactions by applying the THM. The preliminary results of both experiments have been presented recently at the Nuclear Physics in Astrophysics (NPA) VIII Conference, held in Catania last june [46, 47].

3.4. Sub-barrier fusion measurements

Recently, we have also explored the possibility to perform measurements of fusion cross sections at near- and sub-barrier energies. In this case, our facility works as a separator of evaporation

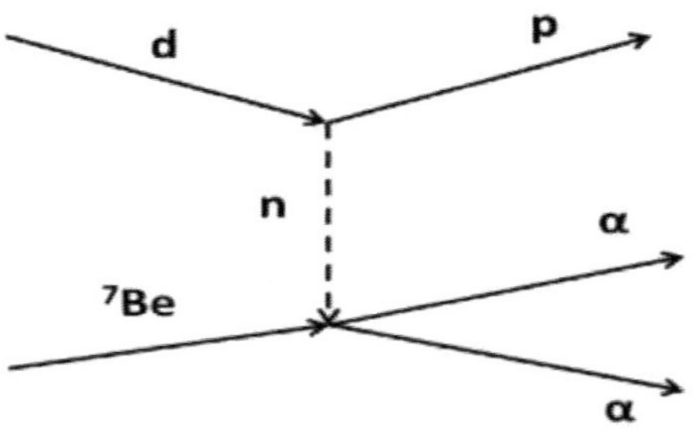

Figure 9. The polar diagram describing the quasi-free ^{7}Be+d interaction. The lower pole describes the ^{7}Be-n interaction of interest.

residues from the incident beam, delivered by the INFN-LNL Tandem XTU accelerator. For this purpose, the facility was slightly modified to allow the transport, identification and detection of the fusion-evaporation residues (ER) at the focal plane of the facility [48]. We measured the fusion reactions for the two systems ^{32}S + ^{48}Ca and ^{32}S + ^{64}Ni where the cross sections are known from previous experiments performed at LNL [49, 50]. The ion optical parameters of EXOTIC were set to maximize the ER yield in the detector system for the various cases. A good and clean separation of the ER from beam-like particles was obtained in the experimental Energy-Time of Flight correlation plots. These encouraging results were compared with the performance of the electrostatic deflector set-up PISOLO, routinely used at LNL for sub-barrier fusion experiments. The beam rejection factor of EXOTIC facility at 0° is comparable to the one of PISOLO deflector at 2° - 3°, while a gain of overall efficiency up to a factor 3 has been obtained with EXOTIC.

4. Conclusion

The EXOTIC facility is dedicated to the in-flight production of low-energy light RIBs and its experimental set-up is fully operational at INFN-LNL. An intense and various experimental program has been performed by using the RIBs delivered by this facility, in the framework of international collaborations.

We performed several experiments aimed to a better comprehension of the interplay between different reaction mechanisms induced by the peculiar features of exotic and weakly-bound nuclei, as shown in section 3.1. In particular, we investigated the reaction dynamics of the ^{7}Be RIB interacting with medium- and heavy-mass targets at Coulomb barrier energies, demonstrating that the ^{7}Be reaction dynamics at Coulomb barrier is not dominated by the breakup process. The detailed analysis of the different channels involved is in progress.

Recently, we extended our interest to the investigation of α-clustering phenomena in light exotic nuclei, employing the TTIK scattering technique with a ^{4}He gas target (section 3.2). We found evidence for α-cluster structure for ^{19}Ne, by using the ^{15}O RIB. This could help to understand the role of this nucleus in the astrophysical ^{15}O(α,γ)^{19}Ne reaction in the break-out from the hot-CNO cycles. Another experiment was performed to study the cluster nature of the ^{15}O nucleus, with the ^{11}C RIB produced at EXOTIC. The analysis is in progress.

An important challenge will be to perform measurements of astrophysical interest with EXOTIC RIBs impinging on solid or gas light targets (section 3.3). In this direction, we will test the production of ^{18}Ne and ^{30}P RIBs, involved in reactions of astrophysical relevance in explosive environments. At the end of 2015, an experiment based on the THM method was performed to investigate the ^{7}Be(n,α)^{4}He reaction to shed light on the cosmological lithium-problem, of interest in the Big Bang Nucleosynthesis. The data analysis is in progress and the preliminary results were presented at the NPA VIII conference held in Catania in June 2017.

Moreover, EXOTIC can be used as a velocity filter to perform fusion-evaporation experiments at near- and sub-barrier energies with stable beams (section 3.4). The feasibility of these experiments has been recently demonstated for the two systems ^{32}S + ^{48}Ca and ^{32}S + ^{64}Ni. In future, EXOTIC could be also used as a velocity filter with the neutron-rich RIBs delivered by the next generation ISOL-type facility SPES (Selective Production of Exotic Species) [51], that now is in construction at INFN-LNL.

5. Acknowledgments
The experimental work presented in this paper was performed with the precious contribution of many people that I would like to warmly aknowledge: A. Boiano, C. Boiano, M. La Commara, G. La Rana, M. Mazzocco, D. Pierroutsakou, C. Signorini, F. Soramel, E. Strano, D. Torresi.

References
[1] Maidikov V Z et al. 2004 *Nucl. Phys.* A **746** 389c
[2] Pierroutsakou D et al. 2007 *Eur. Phys. J. Special Topics* **150** 47
[3] Farinon F et al. 2008 *Nucl. Instr. and Meth.* B **266** 4097
[4] Mazzocco M et al. 2008 *Nucl. Instr. and Meth.* B **266** 4665
[5] Mazzocco M et al. 2013 *Nucl. Instr. and Meth.* B **317** 223
[6] Pierroutsakou D et al. 2016 *Nucl. Instr. and Meth.* A **834** 46
[7] Strano E et al. 2013 *Nucl. Instr. and Meth.* B **317** 223
[8] Signorini C et al. 2010 *Eur. Phys. J.* A **44** 63
[9] Mazzocco M et al. 2010 *Phys. Rev.* C **82** 054604
[10] Mazzocco M et al. 2015 *Phys. Rev.* C **92** 024615
[11] Mazzocco M et al. 2016 *EPJ Web of Conferences* **117** 06006
[12] Mazzocco M et al. 2017 *EPJ Web of Conferences* **163** 00035
[13] Patronis N et al. 2012 *Phys. Rev.* C **85** 024609
[14] Pakou A et al. 2013 *Phys. Rev.* C **87** 014619
[15] Pakou A et al. 2015 *Eur. Phys. J.* A**51** 55
[16] Pakou A et al. 2015 *Eur. Phys. J.* A**51** 90
[17] Strano E et al. 2016 *Phys. Rev.* C **94** 024622
[18] Sgouros O et al. 2016 *Phys. Rev.* C **94** 044623
[19] Sgouros O et al. 2017 *Phys. Rev.* C **95** 054609
[20] Canto L F et al. 2006 *Phys. Rep.* **424** 1
[21] Canto L F et al. 2015 *Phys. Rep.* **596** 1
[22] Liang J Signorini C 2005 *Int. Jour. Mod. Phys.* E **14** 1121
[23] Keeley N et al. 2007 *Prog. Part. Nuc. Phys. Rep.* **59** 579
[24] Keeley N et al. 2009 *Prog. Part. Nuc. Phys. Rep.* **63** 396
[25] Keeley N et al. 2014 *Eur. Phys. J.* A **50** 145
[26] Aguilera E F et al. 2009 *Phys. Rev.* C **79** 021601(R)
[27] Freer M 2007 *Rep. Prog. Phys.* **70** 2149221
[28] Artemov K et al. 1990 *Sov. J. Nucl. Phys.* **52** 408
[29] Fu C et al. 2007 *Phys. Rev.* C **76** 021603(R)
[30] Fu C et al. 2008 *Phys. Rev.* C **77** 064314
[31] Torresi D et al. 2011 *Int. J. Mod. Phys.* E **20** 1026
[32] Wiescher M et al. 2010 *Annu. Rev. Nucl. Part. Sci.* **60** 381
[33] Torresi D et al. 2017 *Phys. Rev.* C **96** 044317
[34] Azuma R E et al. 2010 *Phys. Rev.* C **81** 045805
[35] Mohr P et al. 2014 *Phys. Rev.* C **90** 065806
[36] José J et al. 2001 *Astrophys. J.* **560** 897
[37] José J et al. 2004 *Astrophys. J.* **612** 414
[38] Wrede C et al. 2014 *AIP Advances* **4** 041004
[39] Baur G 1986 *Phys. Lett.* B **178** 135
[40] Spitaleri C et al. 2011 *Phys. of Atomic Nuclei* **74** 1725
[41] Tribble R E et al. 2014 *Rep. Prog. Phys.* **77** 106901
[42] Spitaleri C et al. 2016 *Eur. Phys. J.* A **52** 77

IOP Conf. Series: Journal of Physics: Conf. Series **1014** (2018) 012012 doi:10.1088/1742-6596/1014/1/012012

[43] Broggini C et al. 2012 *J. Cosmol. Astropart. Phys.* **6** 30
[44] Kubono S et al. 2002 *Eur. Phys. J.* A**13** 217
[45] Yanagisawa Y et al. 2005 *Nucl. Instr. and Meth.* A **539** 74
[46] Lamia L et al. 2017 *EPJ Web of Conferences* **165** 01032
[47] Hayakawa S et al. to be published
[48] Strano E et al. 2018 *Nucl. Instr. and Meth.* A **877** 293
[49] Montagnoli G et al. 2013 *Phys. Rev.* C **87** 014611
[50] Stefanini A M et al. 1986 *Nucl. Phys.* A **456** 509
[51] https://web.infn.it/spes/

IOP Conf. Series: Journal of Physics: Conf. Series **1014** (2018) 012013 doi:10.1088/1742-6596/1014/1/012013

Gamma rays as probe of fission and quasi-fission dynamics in the reaction $^{32}S + ^{197}Au$ near the Coulomb barrier

A Pulcini[1], E Vardaci[1], E Kozulin[2], M Ashaduzzaman[1], C Borcea[3], A Bracco[4], S Brambilla[4], S Calinescu[3], F Camera[4], M Ciemala[4], B de Canditiis[1], O Dorvaux[5], I M Harca[2], I Itkis[2], V V Kirakosyan[2], G Knyazheva[2], N Kozulina[2], I V Kolesov[2], G La Rana[1], A Maj[6], I Matea[7], K Novikov[2], C Petrone[3], D Quero[1], P Rath[1], E Saveleva[2], C Schmitt[8], G Sposito[1], O Stezowski[9], W H Trzaska[10] and J Wilson[2]

[1] Università degli Studi di Napoli Federico II - Dipartimento di Fisica "Ettore Pancini", Napoli, Italy
[2] Flerov Laboratory of Nuclear Reactions, Dubna, Russia
[3] Horia Hulubei National Institute of Physics and Nuclear Engineering, Bucharest - Magurele, Romania
[4] I.N.F.N. and Università di Milano, Italy
[5] IPHC, CNRS/IN2P3, Univ. de Strasbourg, France
[6] Institute of Nuclear Physics PAN, Krakow, Poland
[7] IPN, CNRS/IN2P3, Orsay, France
[8] GANIL, CEA/DSMCNRS/IN2P3, Caen, France
[9] Univ. de Lyon, CNRS/IN2P3, IPNL, France
[10] Accelerator Laboratory of University of Jyvaskyla, Finland

E-mail: pulcini@na.infn.it

Abstract. Compound nucleus fission and quasi-fission are both binary decay channels whose common properties make the experimental separation between them difficult. A way to achieve this separation could be to probe the angular momentum of the binary fragments. This can be done detecting gamma rays in coincidence with the two fragments. As a case study, the reaction $^{32}S + ^{197}Au$ near the Coulomb barrier has been performed at the Tandem ALTO facility at IPN ORSAY. ORGAM and PARIS, two different gamma detectors arrays, are coupled with the CORSET detector, a two-arm time-of-flight spectrometer. TOF-TOF data were analyzed to reconstruct the mass-energy distribution of the primary fragments coupled with gamma multiplicity and spectroscopic analysis. Preliminary results of will be shown.

1. Introduction

In reactions where fusion cross-section has to be estimated, it is important to get insight on the possible exit channels. When massive nuclei are involved, at energies around the Coulomb barrier, because of the substantial increase of the Coulomb repulsion, the quasi-fission mechanism (QF) happens to be the most important process counteracting complete fusion (CF) [1]. At the same time shell effects become very important at this energy range [2].

QF is a mechanism in which the composite system separates into two main fragments without forming a compound nucleus. The common properties between QF and compound nucleus fission (FF) make their experimental separation difficult. A few methods have been developed to decompose fission-like fragments into the FF and QF components, mainly based on comparing the experimental properties of the fragments in reactions leading to the formation of the same composite system but having different entrance channel properties. In the lack of a comprehensive theoretical model it is necessary to proceed empirically by searching for correlations between observables and extrapolating the systematics to unknown regions.

Typical pattern of the mass-TKE distribution which suggest the onset of the QF appears as the entrance channel mass symmetry and the reaction coulomb factor $Z_1 \cdot Z_2$ increases. From a mass distribution with a near Gaussian shape, typical of the fusion-fission described by the LDM, the increase of the two parameters above changes the mass distribution of the binary fragments quite remarkably: the mass distribution deviates from the LDM predictions with a more prominent asymmetric component. The average TKE and the variance of the TKE distributions also reflect the same progression.

While in the asymmetric region of the mass distribution it is usually possible to disentangle, to some extent, the component of QF and asymmetric fission modes thanks to the different nature of these two mechanisms, in the symmetric region the two components are overlapped. On one side the LDM cannot predict the relative strength of the CN fission to the total distribution; on the other side QF and FF fragments might have overlapping TKE distributions. The overlap of the two mechanisms in the symmetric mass region constitutes an inescapable problem when CN fission cross section has to be estimated [1].

That being said, it appears clear that a disentanglement of CN fission and QF processes is very hard to achieve and possible only in specific cases. Mass, TKE and angular distributions with their variances are not a good set of observables for the correct disentanglement of the two processes and there isn't still a complete picture on what are the factors which determine the QF process.

In the next section the new approach proposed in this work will be discussed. In section 3 the studied reaction and the experimental setup will be discussed. In section 4 preliminary results will be shown.

2. Gamma ray probe

In the scenario described in the previous section, the search for a possible pathway to disentangle the two process starts from the identification of an additional observable or a new set of observables. The known differences between these two processes can help us finding the way: on one side there is CN fission, a slow process passing through an equilibrium stage; on the other side there is quasi-fission, a faster process with considerable mass transfer and energy dissipation, strongly governed by shell effects. It is reasonable to think that for the slower process the whole orbital angular momentum is transferred into internal degrees of freedom of the compound nucleus and so the two fragments following FF can reach spins higher than those of the fission-like fragments produced by quasi-fission.

A possible way to probe angular momentum of the two fragments is to detect gamma rays. Moreover, with gamma rays it is possible to identify the fragments atomic numbers and to directly investigate shell effects, the physics of the scission point (excitation energy and angular momentum), and the connection between fission modes and feeding of the fragments. Information about angular momentum can be extracted from discrete gamma transitions as well as from gamma multiplicity, M_γ, that is the average number of gamma emitted per event.

Recent studies were performed using modern gamma ray detector arrays and fission decay in reactions on thick targets to stop the recoiling fission fragments. The advantage of this method is that the Doppler broadening is absent if the recoiling fission fragments stop before emitting gamma rays, providing the maximum resolving power for the very dense spectra of gamma ray transitions from hundreds of nuclear species. The disadvantage of this approach is the impossibility to directly correlate gamma rays with fragments mass or TKE. In addition, only states with lifetimes longer than the stopping time for the recoiling fission fragments can be studied.

The drawbacks of the thick-target technique can be overcome by the fragment-gamma coincidence technique, which adds mass selectivity as well as identification from which fission fragment the gamma ray originates. It significantly improves the discrete gamma ray selectivity. The downside is that measurements arising from this technique are affected by gamma rays' energy shift and peak broadening, both due to Doppler effect, so energy corrections are needed [3] for which the measurement of the fragment velocities is required with the highest possible resolution.

Therefore, gating on high angular momentum gamma rays in coincidence with masses and TKE, it should be possible to see that some fragments are more populated in the fission region than in others and this could tentatively lead to a separation of CN fission and QF products even in the symmetric split mass region.

This method of correlating the high angular momentum population with the time scale of a reaction channel can be tested first by using the quasi-elastic channel. If the hypothesis is valid, the gamma transitions measured in coincidence with the quasi-elastic component should come from nuclei populated to lower angular momentum regions and the gamma multiplicity should be smaller than the one in coincidence with the fragments in the symmetric mass region. This would prove the concept and would open the road to experiments to distinguish between QF and CN fission in the symmetric region by employing an additional probe.

3. Experimental setup

To explore the concept described above, the reaction ^{32}S + ^{197}Au, at the energy near the Coulomb barrier, E_{lab} = 166 MeV, was performed at the Tandem ALTO accelerator at IPN Orsay (France). This reaction is characterized by a large fusion-fission cross section, and a negligible contribution from QF. The mass-TKE distribution is therefore characterized by a dominating component from fusion-fission process and the population of high angular momentum regions of the nuclei detected in coincidence with the mass symmetric fragments would therefore be not polluted with components from processes of nearby time scale.

Binary reaction products are detected in coincidence by the two-arm TOF spectrometer CORSET [4]. Each arm consists of a START detector, placed at 60 mm from the target, and a position-sensitive STOP detector, placed at 210 mm from the START. The two arms of CORSET were placed at 68° and 66.5° from the beam axis. These angles were chosen to detect symmetric fragments (the symmetric mass region for this reaction corresponding to 70° in the laboratory frame) and some asymmetric fragments, considering that the angular coverage of each STOP detector is about 10°.

Four beam monitor detectors are placed into the reaction chamber at 18° from the beam axis and at 40 mm from it. The four monitors are placed at angles of 90° with respect to each other.

The prompt gamma rays are detected with ORGAM and PARIS detectors. ORGAM is an array of high-resolution Ge detectors, each surrounded by BGO anti-Compton shields. 10 Ge detectors have been used, all located at backward angles. PARIS is a high efficiency array detector composed by LaBr3 (Ce)-NaI(Tl) phoswich base units. 10 units have been used, 9 packed together plus a single unit, all located at forward angles. The two detector arrays have similar photopeak efficiency in the low energy range, ~1% for ORGAM and ~1.5% for PARIS. ORGAM has better energy resolution (~4 keV at 1408 keV) than PARIS (~60 keV at 1332 keV), however PARIS compensate with a larger dynamical range (up to 20 MeV), compared to ORGAM (up to 2.4 MeV).

START and STOP signals are fed to logic unit modules which generate a trigger if there are two stop signals and at least one start signal. Start and stop signals are used for the measurement of TOFs for the two arms, while coordinates of the hit point on the STOP detectors are obtained as the difference between the time intervals STOP-trigger and coordinate-trigger. Gamma detectors are slave detectors. The readout of the PARIS was accomplished with the LaBrPRO module [5], the rest of the electronic is composed by commercial VME and NIM modules. The energy and time signals of the three setups were all fed into the Vipers [6] data acquisition.

IOP Conf. Series: Journal of Physics: Conf. Series **1014** (2018) 012013 doi:10.1088/1742-6596/1014/1/012013

4. Preliminary results

The ultimate goal for the experimental program is to investigate potential differences in the gamma ray properties depending on the reaction mechanism: fusion-fission or QF. This kind of study does not require a sorting of the data according to fragment masses and TKE, but widely profits from mass gating. In this first stage of the analysis, the goal is to check if it is possible to identify fission-like fragments by their deexcitation gamma rays from the spectra, gating on masses and gamma rays two-fold coincidences and verify that higher angular momenta are populated in the FF region of masses.

4.1. Binary fragments

The velocity vector of a fragment flying in a CORSET arm is obtained computing the flight path of the fragments (in the hypothesis that the fragment originates in the center of the target) and the TOF. Velocity vectors are needed to obtain the mass distribution as well as to perform Doppler correction. Only full momentum transfer binary events were taken into account in this analysis, so the mass of the primary fragments can be obtained using mass and momentum conservation laws. Masses and energies thus obtained need to be corrected due to the energy loss of the fragments along their path towards the detectors. From the TOF-TOF data, the mass-TKE matrix was reconstructed, as shown in figure 1. In the mass-TKE matrix it is possible to distinguish the typical loci for a fusion-fission process. The two lateral parts are elastic and quasi elastic events; the central part is mostly fusion-fission.

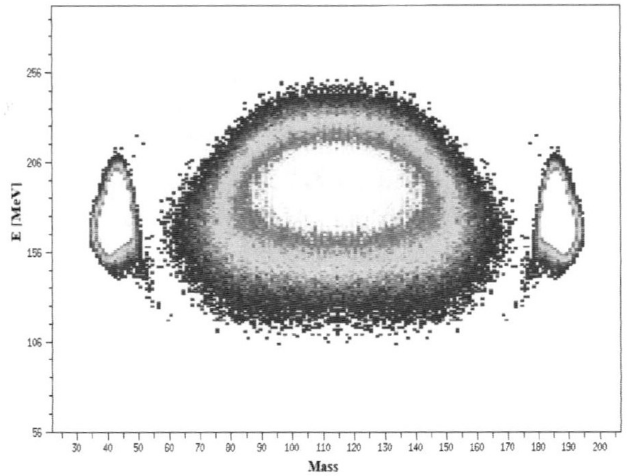

Figure 1. Mass-TKE matrix.

4.2. Gamma rays

From a first gate on the central part of the mass-TKE distribution, gamma energy lines corresponding to target and projectile levels should disappear. This is shown in figure 2. The upper curve represents the total spectrum recorded with the ORGAM array while the lower one represents the total spectrum gated on the fusion-fission part of the mass-TKE distribution. The most visible peaks of the total spectrum, identified as ^{197}Au transitions (figure 2) disappear almost completely in the gated spectrum. Considering the good separation of the two components in the gamma spectrum coming from elastic and quasi-elastic events or CN fission, gamma multiplicities associated to the two regions of the mass-TKE distribution have been compared. Figure 3a shows the gamma rays multiplicity, detected with the PARIS array, for the two different regions: full triangles represent the gamma ray multiplicity of elastic and quasi-elastic events; full circles represent the multiplicity of fusion fission and deep inelastic events. Counts are normalized for comparison and are shown in logarithmic scale. It appears evident that the multiplicity associated with the quasi-elastic events drops faster than in the case of the fusion fission events. Furthermore, the multiplicity extends up to 7 in fusion-fission and up to 1 in quasi elastic. These trends are the proof that a faster process gives rise to less angular momentum in the final fragments. We have to point out that the gamma multiplicity in Fig. 3a is still not corrected for the geometrical efficiency, but the trend is expected to be the same even after applying the correction.

An additional advantage of the present experimental method is that excitation energy regions can be scanned by taking windows of TKE values: a higher TKE value corresponds to less excitation energy available for the fragments (hence, lower multiplicity) and viceversa. Figure 3b shows the average multiplicity versus TKE. The average multiplicity exhibits a clear trend, decreasing as the TKE increases, in agreement with our assumption. This provides an additional gating condition to isolate weak transitions in the sea of the gamma rays measured.

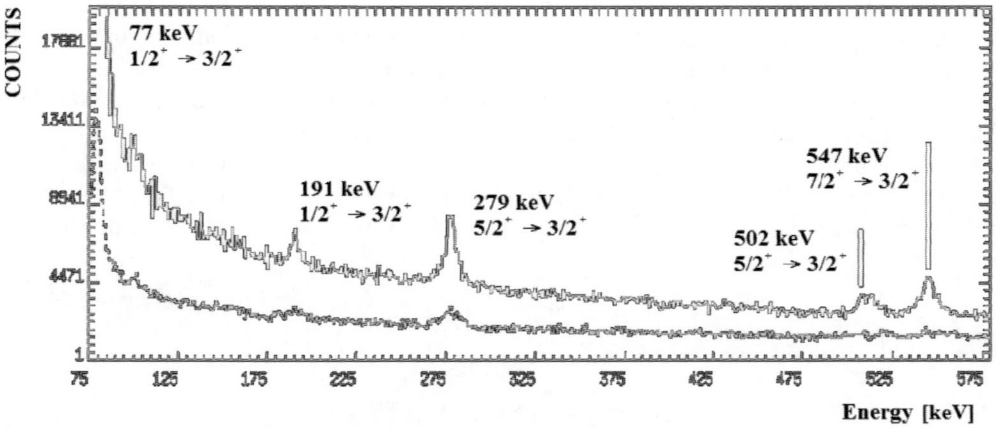

Figure 2. The total spectrum recorded by ORGAM detectors, upper curve, compared with the spectrum relative to the central part of the mass-TKE distribution, lower. The ^{197}Au transitions identified in the total spectrum are marked with their energy and angular momentum.

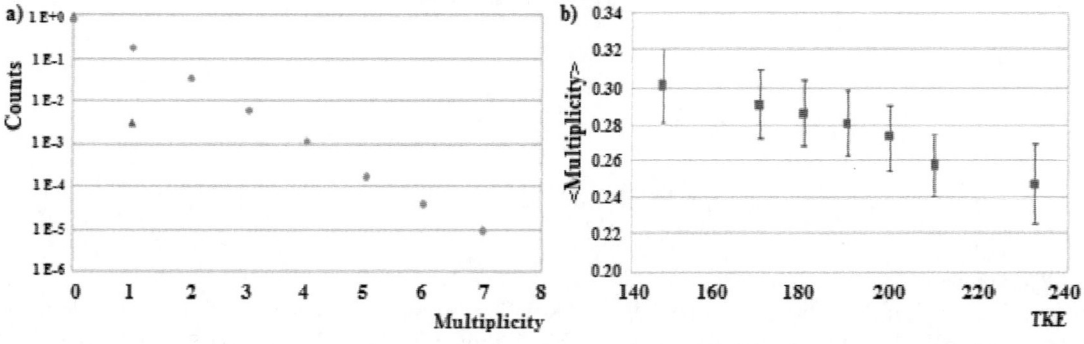

Figure 3. a) Multiplicity distributions for elastic and quasi-elastic channel (full triangles) and for fusion-fission channel (full circles); b) Average gamma multiplicity in coincidence with fission fragments versus their TKE.

A further prove in support of this conclusion could be to find gamma transitions between levels with high angular momentum in coincidence with masses in fusion-fission region. For this kind of analysis Doppler correction is mandatory. The energies of the gamma rays have to be corrected for both fission fragments, being impossible to know in advance from which fragment a gamma ray originates. Therefore, two gamma spectra are produced from one gamma spectrum after Doppler correction and must be analyzed separately [7]. An entire set of gamma transitions have been found for the nucleus ^{129}La connecting the level with $E^* = 5934.2$ keV and $J_\pi = 39/2+$ to the level with $E^* = 3420.6$ keV and $J_\pi = 27/2+$. Figure 4 shows the gamma ray energy spectra detected with the ORGAM array, in coincidence with both the mass (130 ± 3 a.m.u.) and one low lying transition (270 ± 5 keV) of the ^{129}La isotope ($E_{\gamma 1}$=269.7 keV), Doppler-corrected for the ^{129}La fragment. Although statistics precludes unambiguous identification of the gamma transitions, these results and the previous ones confirm the expectation that in the FF process higher angular momentum levels can be produced. Consequently, we can support the idea that processes with longer time scales can convert much more orbital angular momentum into the fragments. The different regions of angular momentum populated in FF and QE, at least for the ions selected so far, confirms this conclusion. On the same foot, we can

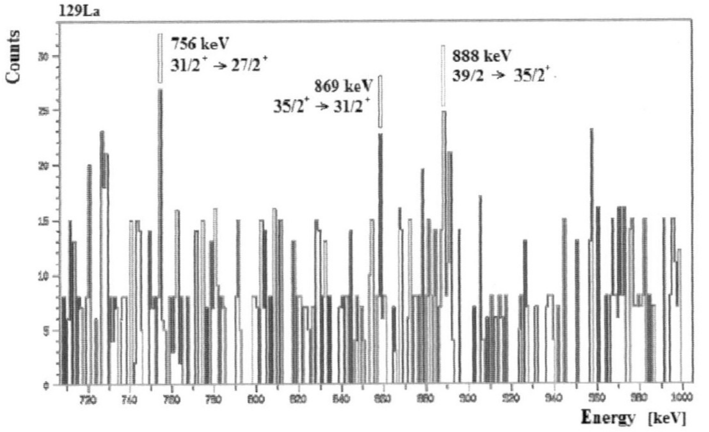

Figure 4. Some high angular momentum transition gamma rays of ^{129}La.

reasonably expect the same effect also in QF and FF. Hence, these results give support to our initial expectations.

As a final comment we observe that, ^{129}La is a proton rich nucleus, and thus the partner must be the neutron rich nucleus ^{100}Sr whose level scheme is only fairly known. This means that looking for more proton rich nuclei (for instance the partner of the well-known ^{127}La is the unknown ^{102}Sr), with FF or QF reaction channels it might be possible to populate unknown neutron-rich nuclei and that the reconstruction of their level scheme can be pursued by using fragment-fragment coincidences. This is a very important result because there is no other known mean to produce such nuclei. At the same time, it is clear that these preliminary data need further processing to identify other possible unknown isotopes.

5. Conclusions

The preliminary results discussed in this work confirm that the gamma probe can play a very important role in disentangling FF and QF in the regions of the mass-TKE matrix where they are overlapped. Figure 3 is the key result of this work and shows how, gating on high multiplicity processes, favors the selection of slower processes. However, we must stress that these are preliminary results. The multiplicity distributions need to be corrected for the geometrical efficiency. For instance, even though only three gamma rays are detected in an event, they may come from events of larger multiplicity. Mass-TKE distributions also need correction for the energy losses: these corrections will improve the mass-TKE distribution and, most of all, the energy resolution of gamma energy spectra, affected by the errors introduced in the Doppler corrections. In any case, all of the mentioned corrections do not affect the main result of this work.

Yet, also high angular momentum spectroscopy can be performed by populating nuclei in level regions not accessible with usual fusion-evaporation reactions. As a by-product of this analysis, we have discovered that neutron-rich (possibly unknown) nuclei can be populated as partners of proton-rich nuclei. This means that a completely new spectroscopy can be accessible by using fragment-fragment coincidences. Still, the analysis is only at the beginning of a long process. After the mass-

TKE distributions correction for the energy losses, it would be very important to narrow the window of masses to reduce the gamma background and to look for unknown nuclei.

For the future plans, it is crucial to test this method in condition of major interest: when FF and QF are overlapped with the same intensity in the same mass region. It would also be important to benefit of these results and take advantage of the properties of the QF and FF to populate unknown neutron-rich nuclei. Reactions can be chosen carefully to populate neutron-rich regions of the nuclide chart of specific interest, like the one of interest from the r-process. However, it is clearly evident that the experimental condition must be kept at the optimum and that much larger gamma ray efficiency is necessary to reach a sufficient statistics.

6. Acknowledgement

This work was supported by FLNR, JINR, INFN, the IN2P3-JINR Agreement No 14-90, the ENSAR2 project "N-SI-88", by the financial contribution of Romania to the Joint Institute of Nuclear Research, Dubna, Russia to the project no. 03-5-1094-2010/2016, by the Polish National Science Centre under Contract No. 2013/08/M/ST2/00257 and the French LEA COPIGAL project.

References

[1] Itkis M G, Vardaci E, Itkis I M, Knyazheva G N and Kozulin E M 2015 *Nucl. Phys.* A **944** 204
[2] Kozulin E M *et al.* 2014 *Phys. Rev.* C **89** 014614
[3] Aiche M *et al.* 2003 *Nucl. Inst. and Met. in Phys. Res.* A **508** 367-377
[4] Kozulin E M *et al.* 2008 *Instr. and Exp. Tec.* **51**
[5] Boiano C, Brambilla S, Riboldi S, Giaz A and Camera F 2015 *NSSMIC, IEEE, N2AP* **43**
[6] Romoli M, Di Pietro M, Vardaci E *et al.* 2004 *NSS, IEEE* **3** 1932-1936
[7] Kozulin E M, Vardaci E *et al.* 2016 *Eur. Phys. J.* A **52** 293

International Workshop "Nuclear Reactions on Nucleons and Nuclei" IOP Publishing
IOP Conf. Series: Journal of Physics: Conf. Series **1014** (2018) 012014 doi:10.1088/1742-6596/1014/1/012014

Study of resonances produced in Heavy Ion Collisions

L. Quattrocchi[1,2], L. Acosta[2,3], L. Auditore[1,5], G. Cardella[1], A. Chbihi[6], E. De Filippo[1], F. Favela [1], B. Gnoffo [1], G. Lanzalone[4,9], I. Martel[7], N. S. Martorana[2,4], A. Pagano[1], E.V. Pagano[2,4], M. Papa[1], S. Pirrone[1], G. Politi[1,2], F. Porto[2,4], F. Rizzo[2,4], P. Russotto[1], A. Trifirò[1,5], M. Trimarchi[1,5], G. Verde[1,8], M. Veselsky[10]

[1] INFN, Sezione di Catania, Catania, Italy
[2] Dip. di Fisica e Astronomia, Università di Catania, Catania, Italy
[3] Instituto de Fisica, Universidad Nacional Autonoma de Mexico, Mexico
[4] INFN, Laboratori Nazionali del Sud, Catania, Italy
[5] Dipartimento di Scienze MIFT, Università di Messina, Messina, Italy
[6] GANIL,CEA-IN2P3-CNRS, Caen, France
[7] Departamento de Física Aplicada, Universidad de Huelva, Huelva, Spain
[8] IPN Orsay, Orsay, France
[9] Università Kore, Enna, Italy
[10] Slovak Academy of Sciences, Bratislava, Slovakia

E-mail: lquattrocchi@unime.it

Abstract. At Laboratori Nazionali del Sud of Catania an experiment has been carried out in order to investigate the correlations between particles produced in $^{12}C+^{24}Mg$ reaction at 35 AMeV incident energy. Two α correlation has been explored because provide information about temperature of ^{8}Be nuclei produced in the reaction, while three α correaltion has been studied in order to evaluate the competition between sequential and direct decay mode of resonances produced in ^{12}C quasi-projectiles.

1. Introduction

In the framework of heavy ion collisions two- and multi-particles correlations are a powerful tool to probe the properties of emitting sources produced and thus involved dynamics [1] but also to explore certain structure properties of unbound states such as the spins [2]. Correlations between light particles can also be used to study the population of unbound states of decaying parent nuclei [3, 4]. Population probability of two internal states, Y_1 and Y_2, separated by an energy ΔE, is determined by the temperature T of the system according to Boltzmann factor $e^{(\Delta E/T)}$ [5]. If these states decay into specific particle pairs, they can be isolated by means of the two-particle coincidence spectra and their population can be estimated and linked to emission temperature of the decaying system. But these techniques have been also used in experiment performed with 4π detectors in order to disentangle direct and sequential decay modes of the produced resonances and to estimate their relative contribution to the decay width. For example three-α correlation has been explored to study the decay mechanisms of ^{12}C excited states [6, 7]. With the aim to investigate dynamics and also to extract nuclear structure properties of produced resonances in heavy ion collisions, correlations between two-

and multi-particle emitted in peripheral ^{12}C+^{24}Mg reaction at 35 AMeV incident energy have been analyzed.

2. Experimental results

The studied experiment has been carried out at Laboratori Nazionali del Sud of Catania; detection and identification of charged reaction products have been possible using the forward part of CHIMERA 4π multi-detector [8]. The two particle correlation function, $1 + R(x_1, x_2)$, can be defined as:

$$1 + R(x_1, x_2) = C_{12}\,\frac{Y_{12}(x_1, x_2)}{Y_1(x_1)\cdot Y_2(x_2)}, \tag{1}$$

where $Y_{12}(x_1, x_2)$ is the two-particle coincidence yield, while $Y_1(x_1)\cdot Y_2(x_2)$ is the product of the single particle yields and C_{12} is a normalization constant. In Eq. 1 x_1 and x_2 represent generic variables chosen to express the two bodies correlation. Traditionally, for correlation between two particles, momentum of relative motion, q_{rel}, is used. In Fig. 1 α-α correlation function is reported in term of relative momentum (q_{rel}). The $Y_{12}(x_1, x_2)$ coincidence yield has been obtained using all α-α couples detected in coincidence in the same event while the denominator of Eq. 1, $Y_1(x_1)\cdot Y_2(x_2)$, has often been approximated with the uncorrelated two-particle yields $Y_{12}^{unco}(x_1, x_2)$ and may be constructed using different techniques. In this work the emission of uncorrelated α-α pairs are simulated by extracting randomly their energy, polar (θ) and azimuthal (Φ) angles, from single particle yield spectra. In this way it is possible to construct the relative momentum of uncorrelated α-α pairs.

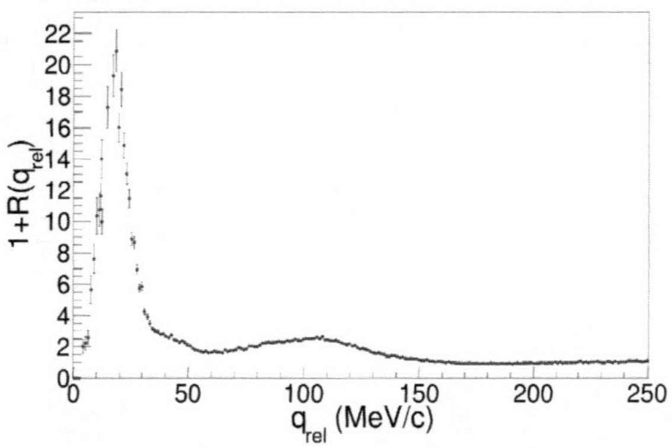

Figure 1. Experimental α-α correlation function versus momentum of relative motion, q_{rel}, measured in ^{12}C+^{24}Mg reaction at 35 AMeV.

The correlation function on Fig. 1 shows a pronounced peak centered around q_{rel} $\approx$18.5 MeV/c corresponding to decay of the unstable ground state of ^{8}Be (J^{π}=0^{+}, Γ=6.8 MeV). The second broader peak at q_{rel} $\approx$105 MeV/c corresponds to the decay of its first excited state at E_{ex}(^{8}Be)=3.04 MeV (J^{π}=2+, Γ=1.5 MeV). This α-α correlation function can be used to extract information about the emission temperature of ^{8}Be nuclei produced in ^{12}C+^{24}Mg reaction. Instead according to the thermal model the primary relative population of particle unstable states (i, j; $E_i<E_j$) is strictly related to the temperature T of the system, according to Boltzmann factor, $e^{-\Delta E/T}$ [9]:

$$R_{i,j} = \frac{Y_i}{Y_j} = \frac{(2J_i + 1)}{(2J_j + 1)}e^{-\Delta E/T}. \tag{2}$$

IOP Conf. Series: Journal of Physics: Conf. Series **1014** (2018) 012014 doi:10.1088/1742-6596/1014/1/012014

With the aim of estimating the temperature associated to the decay of ^{8}Be nuclei, experimental α-α coincidence yield is compared to results of Monte-Carlo simulations.

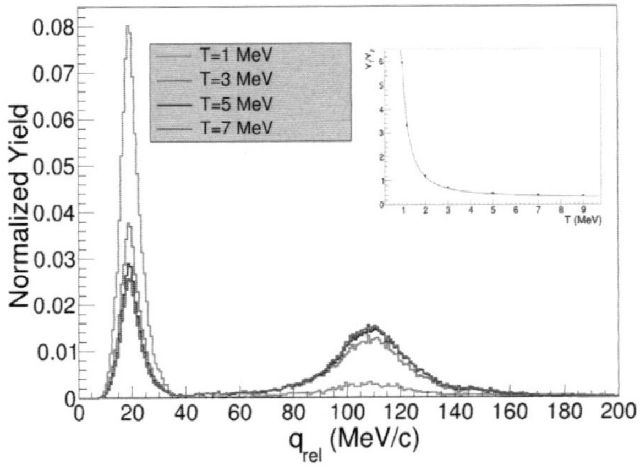

Figure 2. Simulated α-α coincidence yields as a function of relative momentum q_{rel} obtained using different temperatures. The relative population of the two states has been simulated according to Eq. 2. The insert shows Y_1/Y_2 predicted by simulations as a function of temperature.

The decay ^{8}Be$\rightarrow 2\alpha$ has been simulated taking into account the temperature dependence of relative population of parent nucleus states shown in Eq. 2. Fig 2 reports simulated α-α coincidence spectra as a function of q_{rel} at different temperatures (1, 3, 5, 7 MeV). Simulated data have been filtered through the geometry and detector response of CHIMERA. Yield ratio Y_1/Y_2 predicted by Monte Carlo simulations as a function of temperature is reported in the insert of Fig. 2 where Y_1 and Y_2 denote integrated yields of ^{8}Be at lower and higher excitation energy, respectively. The populations of ^{8}Be unstable states are extracted by integrating the experimental α-α coincidence yield over the range of energies dominated by the corresponding resonances. The comparison with simulated events suggests a value of temperature of about 5 MeV. Such value is in reasonable agreement with emission temperature measurements at similar beam energies.

Similarly to two-particle correlation function, correlation of three or more particles can be constructed in heavy ion reactions. In the present work three-α correlations have been explored with the aim of studying the competition between different decay processes of resonances produced in ^{12}C. In order to isolate α particles resulting from decay of excited ^{12}C quasi-projectiles, a selection based on the velocities of fragments has been applied. In particular the parallel velocity reconstructed from the center of mass of three α particles detected in the same event, has been required to be greater than 80% of the beam velocity (8 cm/ns). This criterion has been suggested by comparison with results of simulations performed with HIPSE model calculations [10]. The 3 α correlation function is shown on Fig. 3 and it is reported as a function of excitation energy $E_{ex}=E_{tot}-Q$, where E_{tot} is the total kinetic energy of the three α particles in the center of mass reference frame and Q is the Q-value for the corresponding decay channel. It shows two peaks: the first one, centered around E_{ex}=7.74 MeV, corresponds to the Hoyle state (E_{th}=7.65 MeV, Γ=8.5 eV); the second one, centered at E_{ex}=9.8 MeV, arises from the overlap of states at E_{ex}=9.64 MeV (3^-), E_{ex}=10.3 MeV (0^-) and possibly at E_{ex}=9.7 MeV (2^+).

The uncorrelated background has been built with a similar procedure used to construct two-

IOP Conf. Series: Journal of Physics: Conf. Series **1014** (2018) 012014 doi:10.1088/1742-6596/1014/1/012014

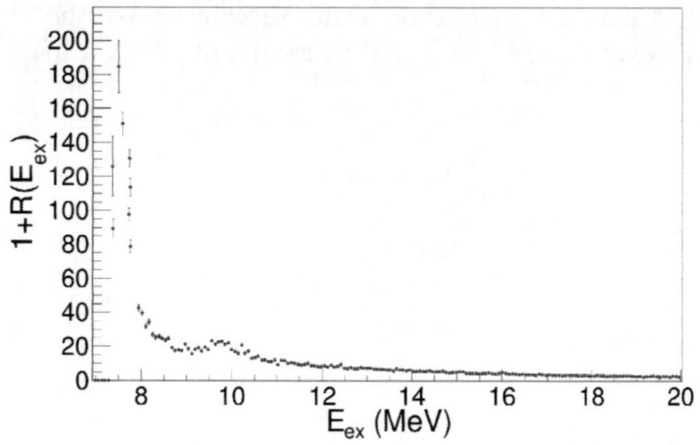

Figure 3. 3 α correlation function versus excitation energy of ^{12}C obtained in ^{12}C+^{24}Mg reaction at 35 AMeV.

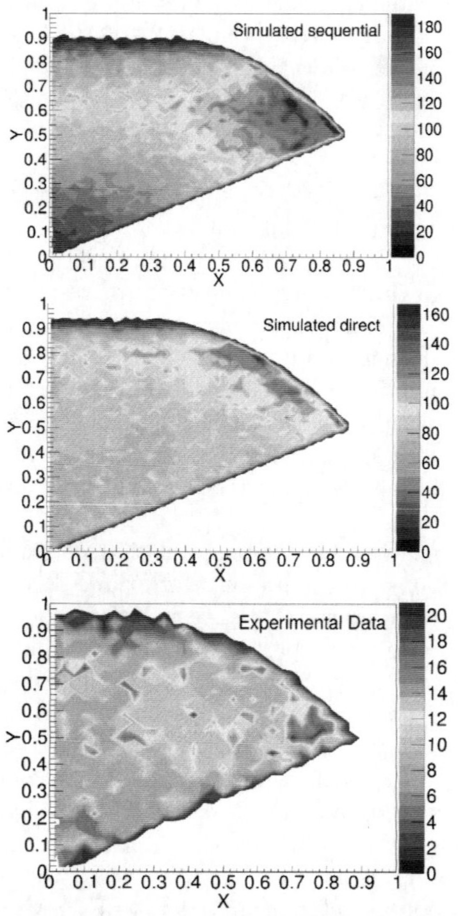

Figure 4. Dalitz plots of simulated events for sequential (top panel) and direct phase-space (central panel) decay of Hoyle state. Bottom panel: Dalitz plot, corresponding to decay from Hoyle state region, constructed with experimental data collected in ^{12}C+ ^{24}Mg reaction.

IOP Conf. Series: Journal of Physics: Conf. Series **1014** (2018) 012014 doi:10.1088/1742-6596/1014/1/012014

particle one, described above; in particular α particles have been chosen randomly by single-particle spectra. In order to study decay mechanisms of the observed resonances, possibly disentangling sequential decays from direct three-body ones, a procedure based on construction of Dalitz plots has been developed. There are several couples of variables (x,y) used to construct this kind of plot; here the following coordinates have been used [11]:

$$X = \sqrt{3}(\epsilon_j - \epsilon_k), \qquad Y = 2\epsilon_i - \epsilon_j - \epsilon_k, \tag{3}$$

where $\epsilon_{i,j,k} = E_{i,j,k} / (E_i+E_j+E_k)$ are the energies of particles in the center of mass reference frame, normalized to total decay energy. In order to interpretate experimental results, dedicated Montecarlo simulations in which ^{12}C decays in 3 α particles directly ($^{12}\mathrm{C} \rightarrow \alpha-\alpha-\alpha$) or passing through the ground state of ^{8}Be ($^{12}\mathrm{C} \rightarrow {}^{8}\mathrm{Be}\text{-}\alpha \rightarrow 3\alpha$) have been performed. Simulated data have been processed by a filter that take into account all effects induced by the used apparatus.

Fig. 4 shows Dalitz plots for simulated events corresponding to sequential mechanism (top panel) and direct 3 α decay according to phase-space equiprobability DDΦ (central panel) of the Hoyle state. As it emerges from simulated events, DDΦ results in a flat distribution, while in the case of sequential decay (SD) the events are distributed in a narrow horizontal band of uniform intensity. The plot constructed with experimental data, reported on bottom panel of Fig. 4, exhibits a more uniform distribution that does not allow us to exclude any of the two decay mechanisms. For better evaluations, monodimensional ϵ_i distributions are used. ϵ_i is the highest normalized energy among those of the 3 α particles emitted from the region around the Hoyle state, and it gives information about decay processes [13].

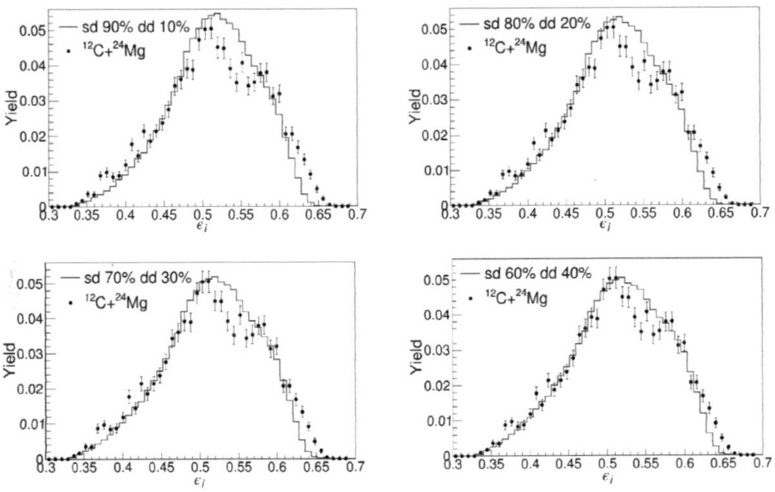

Figure 5. ϵ_i distribution constructed with data collected in ^{12}C+^{24}Mg reaction fitted with four combinations of percentages related to sequential and direct decay mechanisms.

In the case of sequential mechanism, ϵ_i distribution should have a peak around 0.5 while in direct processes ϵ_i should range between 1/3 and 2/3, corresponding to the case of three α particles with an equal energy, and the case in which α particle is emitted in a direction opposite to that of the other two, respectively. In order to obtain a quantitative estimation for the probability of each individual sequential and direct decay mechanism, experimental distributions have been compared to simulated events in which different percentages of direct and sequential decays have been introduced. Fig. 5 shows the results of such comparison of calculations to experimental data in ^{12}C+ ^{24}Mg reaction. As clear from the fits, agreement between simulated and experimental data improves when increasing the percentage of direct decay component. It is very difficult to extract the percentage of direct and sequential components from this comparison because of limited energy and angular resolution. Nevertheless the results reported in Fig. 5 evidence the presence of an important direct decay component that seems lower than 60% but also larger than 20% of the total decay width.

3. Conclusions

In the present work a study of two- and multi-particle correlations in ^{12}C+^{24}Mg reaction has been carried out using CHIMERA 4π multi-detector. A method to extract the emission temperature of excited systems from two-particle coincidence spectra has been discussed. This analysis has provided an emission temperature of ^{8}Be nuclei that is compatible with values obtained in this kind of reactions. Instead three α correlation has been investigated to study the decay processes of Hoyle state produced in ^{12}C quasi-projectiles. From the comparison between experimental and simulated data the contribution of a direct mechanism component emerge in the decay of this analyzed resonance according to other works performed with heavy ion reactions [6, 7]. The quantitative contributions of various 3α decay mechanisms of the Hoyle state have been determined in many works. A large part of these studies, performed with direct reactions [15, 14, 16] or inelastic scattering [13] evidence a very low direct contribution in the decay of this state. These discrepancies could be due to different mechanisms that dominate these reactions. It is possible that the observed effects are influenced by the medium in which the resonance is originated and decays; infact heavy ion collisions are characterized by more dissipative mechanisms and by longer interaction times compared to direct reactions. In order to clarify this intriguing topic it will be necessary to perform new experiments with different reaction systems at various dissipation degrees.

References

[1] Verde G *et al.* 2006 *Eur. Phys. J. A* **30** 81-108.
[2] Tan W P *et al. Phys. Rev. C* **69** 061304.
[3] Xi H F *et al.* 1998 *Phys. Rev. C* **58** R2636.
[4] Serfling V *et al.* 1998 *Phys. Rev. Lett.* **80** 3928.
[5] Morrissey D J *et al.* 1984 *Phys. Lett. B.* **148** 423.
[6] Raduta Ad R *et al.* 2011 *Phys. Lett. B* **705** 65.
[7] Grenier F *et al.* 2008 *Nucl. Phys. A* **811** 233.
[8] Pagano A *et al.* 2004 *Nucl. Phys. A* **734** 504.
[9] Morrissey D J, Benenson W., and Friedman W. A. 1994 *Annu. Rev. Nucl. Part. Sci.* **44** 27.
[10] Lacroix D *et al.* 2004 *Phys. Rev. C* **69** 054604.
[11] Kirsebom O *et al.* 2012 *Phys. Rev. Lett.* **108** 202501.
[12] Dalitz R H 1953 *Phil. Mag.* **44** 1068.
[13] Itoh M *et al.* 2014 *Phys. Rev. Lett.* **113** 102501.
[14] Dell'Aquila D *et al.* 2017 *Phys. Rev. Lett.* **119** 132501.
[15] Freer M *et al.* 1994 *Phys. Rev. C* **49** R1751.
[16] Rana T K *et al.* 2013 *Phys. Rev. C* **88** 021601.

IOP Conf. Series: Journal of Physics: Conf. Series **1014** (2018) 012015 doi:10.1088/1742-6596/1014/1/012015

Production of n-rich Nuclei along the Closed Shell N=126 in the collision ^{136}Xe + ^{208}Pb @E_{lab}=870 MeV

D Quero[1], E Vardaci[1], E M Kozulin[2], V A Zagrebaev[2], L Corradi[3], A Pulcini[1], G La Rana[1], I M Itkis[2], G N Knyazheva[2], K Novikov[2], I Harca[2], E Fioretto[3], A M Stefanini[3], D Montanari[3], G Montagnoli[3], F Scarlassara[3], S Szilner[4], T Mijatović[4] and W H Trzaska[5]

[1]Università degli Studi di Napoli Federico II - Dip. di Fisica *"Ettore Pancini"* and INFN, Napoli, Italy

[2]Flerov Laboratory of Nuclear Reactions, Joint Institute for Nuclear Research, Dubna, Russia

[3]Istituto Nazionale di Fisica Nucleare, Laboratori Nazionali di Legnaro, Legnaro (PD), Italy

[4]Ruder Bošković Institute, and Univeristy of Zagreb, Croatia

[5]Department of Physics, University of Jyväskylä, FI-40351, Jyväskylä, Finland.

E-mail: quero@na.infn.it, vardaci@na.infn.it

Abstract. Multi-nucleon transfer reactions are nowadays the only known mean to produce neutron-rich nuclei in the *Terra Incognita*. The closed-shell region N=126 is crucial for both studying shell-quenching in exotic nuclei and the r-process, being its last "waiting-point". The choice of suitable reactions is challenging and a favorable case is ^{136}Xe+^{208}Pb, near the Coulomb barrier, because their neutron shell-closures play a stabilizing role, favoring the proton-transfer from lead to xenon. TOF-TOF data were analyzed to reconstruct the mass-energy distribution of the primary fragments. Preliminary results of an experiment held at Laboratori Nazionali di Legnaro with PRISMA, aimed at A and Z identification of the products, will be shown.

1. Introduction

The exploration of the the unknown regions of the Segré map and the potential discovery of the *Island of Stability* represents one of the most interesting challenges in nuclear physics and astrophysics. In order to reach these goals, the production of super-heavy species and half-lives and masses measurement are extremely important and would contribute to the discussion about the quenching of shell effects in nuclei with large neutron excess. Fusion reactions with neutron-rich radioactive nuclei and neutron capture are unfeasible channels, because of the low intensity of the currently available Radioactive Ion Beams and the insufficient neutron fluxes from working nuclear reactors. Multinucleon transfer reactions and *Quasi-Fission* process, employing stable beams, look more feasible, with cross sections of the order of mb or μb. Emblematic is the collision ^{136}Xe+^{208}Pb at energies close to the Coulomb barrier. In this case, both nuclei have a closed neutron shell (N=82 and N=126) and, thus, proton transfer from Pb to Xe might be favorable, allowing the exploration of the closed shell N=126, corresponding to the last known "waiting point" of the r-process. The Q-value distribution around zero for such transfers would result in slightly excited fragments, having in this way a greater survival probability. M-TKE distributions of the binary products can be measured using simple kinematic analysis (TOF-TOF data). The results of a preliminary experiment [1] show possible transfers up to about 20 nucleons. Furthermore, there is a good agreement with cross section calculations in the region

IOP Conf. Series: Journal of Physics: Conf. Series **1014** (2018) 012015 doi:10.1088/1742-6596/1014/1/012015

of interest (A~200, Qgg~0) and the yield is even underestimated [2] by a factor 2 in the region of asymmetric mass division.

2. Production of Nuclear Species in the *Terra Incognita*

To fill the "blank region" in the Segré map, the multi-nucleon transfer process in low-energy collisions of heavy ions has been proposed [2]. It is well known that in *Deep-Inelastic* collisions of heavy ions the energy of relative motion is quickly transformed into the internal excitation of the fragments which afterwards de-excite by evaporation of light particles (mostly neutrons). At a first glance, this does not seem of any help for the production of neutron-rich nuclei. However, if the colliding energy and the reaction Q-value are rather low, the primary fragments might not be very much excited and will descend to their ground states after evaporation of few neutrons, thus remaining far from the stability line. Moreover, it has been shown experimentally that even at low bombarding energies (close to the Coulomb barrier) the cross sections for transfer of several protons and neutrons are still rather high [3], and this kind of reactions could be considered as an alternative way for the production of exotic nuclei.

The satisfactory agreement between experimental data and calculation [2], within the model [5] proposed by Zagrebaev and Greiner, gave confidence in obtaining reliable cross section estimations of near-barrier multinucleon transfer reactions producing heavy neutron-rich species. It is clear that an appropriate choice of colliding partners is important for the production in a specific region and shell-closures effects play a very important role in the mass rearrangement both in *Fusion-Fission* and DI processes, possibly inducing the production of neutron-rich nuclei.

3. Predictions of the Dynamical Model

This semi-classical model proposes a dynamical description of a nuclear reaction depending only on some *bulk* degrees of freedom (*DoF*) which can provide a realistic picture of collective behavior and mass transfer. All the remaining *DoF* are treated as a heat bath. The equations (Langevin, eq. 1) describe the time evolution in terms of the variables and their conjugate momenta. The mechanism for transferring energy between the collective degrees of freedom and the heat bath is dissipation by viscosity and a stochastic term is used to take into account fluctuations.

$$\mu\ddot{q} = -\frac{\partial V}{\partial q} - \gamma\dot{q} + \sqrt{\gamma T}\,\Gamma(t). \tag{1}$$

The choice of proper *DoF*s is crucial to describe simultaneously several competitive mechanisms such as DI and QF and estimate the cross sections. The other key ingredient of the model is clearly the potential energy depending on the chosen set of degrees of freedom.

There is a link between the reaction channel, shell-effects and observables such as the fragment mass and Total Kinetic Energy (Fig. 1). The potential energy (often referred to as "potential landscape") shows valleys corresponding to a higher binding energy, therefore to shell closures. These valleys are related to mass divisions that can be associated to specific *loci* in the M-TKE distribution and specific reaction mechanisms, showing how these two observables can be of paramount importance from the experimental point of view.

It is interesting to note that QF may lead to an almost symmetric mass division that would be indistinguishable, in the M-TKE distribution, from the symmetric mass division that can arise from FF. In order to disentangle the two mechanisms it is necessary to look at other observables, such as gamma emission.

An example of calculation conducted within this model is shown in Fig. 2, a M-TKE distribution of fragments from the reaction ^{238}U+^{248}Cm. It is noticeable the possibility of asymmetrizing massive transfer, from light to heavy partner, leading to super-heavy nuclei: the masses inside the circle in figure are in a region close to the center of the Island of Stability.

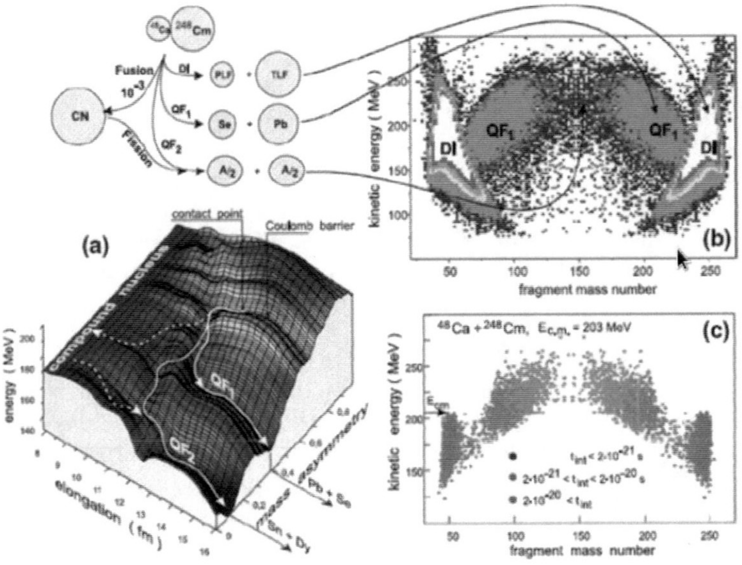

Figure 1. a) the solid lines show the QF trajectories going to the Pb and Sn valleys in the potential landscape. The dashed curves correspond to FF. Experimental b)and calculated c) M-TKE distribution of reaction fragments in collision at 203 MeV center-of-mass energy. Different colors in c) indicate the interaction time of different events. QF and FF cannot be disentangled by means of just mass and TKE (from [4, 5])

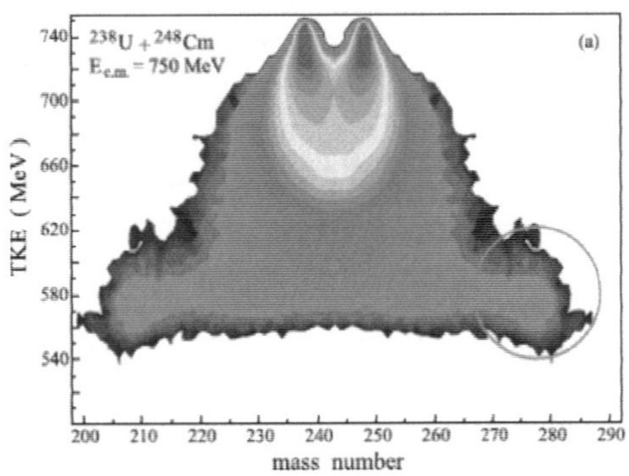

Figure 2. M-TKE distribution of ^{238}U+^{248}Cm, calculated within the model [4]. Asymmetrizing transfers result in super-heavy products. The circle corresponds the center of the Island of Stability.

4. Xe+Pb Reaction and the Closed-Shell N=126

The basic idea is to exploit stable beams and start from the case of the closed shell N=126. The study of the multinucleon transfer at low energies in the reaction ^{136}Xe+^{208}Pb assumes a great importance: if the model is able to predict, at a good extent, the cross sections, there will be more confidence in searching for superheavy element formation in the multinucleon transfer between massive nuclei and this mechanism would prove to be an important candidate for the production of elements, not reacheable by other known mechanisms. The choice falls on xenon

IOP Conf. Series: Journal of Physics: Conf. Series **1014** (2018) 012015 doi:10.1088/1742-6596/1014/1/012015

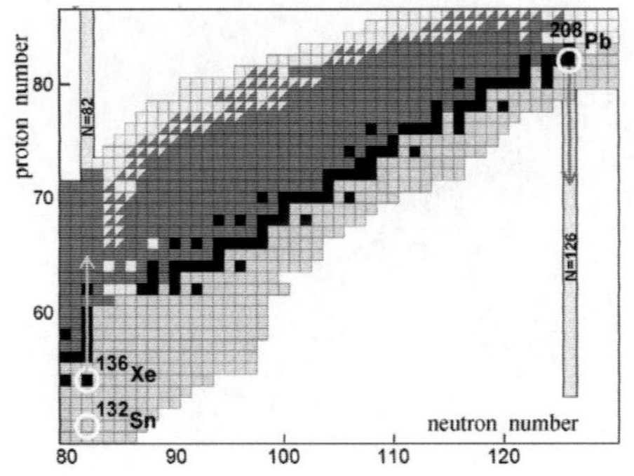

Figure 3. The favorable proton transfer from Pb to Xe allows the exploration of the region N=126. The TLF would slide into the Terra Incognita, keeping its neutron core: an example may be ^{202}Os [5].

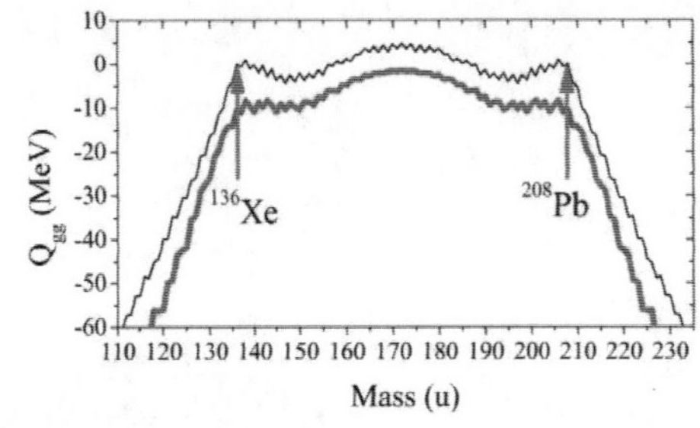

Figure 4. Q-value distribution for transfers in ^{136}Xe+^{208}Pb. In red the mean values, in black the maximum ones [1]. Values around zero for symmetric configurations and negative for asymmetric ones.

and lead because of their neutron shell-closure (N=82 and N=126): both have a closed shell neutron core and, if this stabilizing effect persists while descending into the *Terra Incognita*, the proton transfer from target to projectile could be favorable [1]. It could allow the exploration of the region with N=126 as shown in Fig. 3. An example of produceable species may be ^{202}Os [5].

The Q-value distribution (Fig. 4), which averages around zero in the symmetric mass region [2, 1] and the low bombarding energies may result in poorly exited fragments with greater survival probability against neutron evaporation or fission. Since the Total Kinetic Energy Loss and excitation energy are given by

$$TKEL = E_{cm} - TKE \quad (2)$$
$$E^* = TKEL + Q_{gg}, \quad (3)$$

and being $Q_{gg} \sim 0$, the TKEL trend equals to a good degree the excitation of the primary fragments. The bombarding energy of 870 MeV, near the Coulomb barrier ($V_C = 421.5\,MeV$, in Bass parametrization [7]) would then keep low the excitation of the fragments.

5. Mass Reconstruction

To extract the requested mass-TKE distribution of primary fragments, a time of flight spectrometer is used. By measuring the TOF of two fragments in coincidence and their position on stop detectors, their velocity vector can be evaluated, assuming that it would not be affect heavily by particle evaporation. The application of conservation laws such as energy (eq. 2, 3), momentum (5) and mass number (4) allows the kinematic reconstruction of the event and, in particular, the observables of interest.

$$M_{tar} + M_{proj} = M_1 + M_2 \quad (4)$$
$$M_{proj}\vec{V_{proj}} = M_1\vec{V_1} + M_2\vec{V_2} \quad (5)$$

The experimental apparatus used to study the reaction ^{136}Xe+^{208}Pb @LNL consists of the magnetic spectrometer *PRISMA* coupled with a *Bragg chamber* detector. Fig. 5 shows a schematic drawing of the two arms placed at 45° and 52°, respectively.

The TOF-TOF measurements are taken at the position-sensitive entrance detectors, a MCP in PRISMA arm and a PPAC in Bragg arm, used for full kinematic reconstruction: evaluation of angles, flight paths, velocities and, eventually, mass and energy.

The atomic number of one of the fragments can be evaluated with the use of a $\Delta E - E$ telescope-like ionization chamber at the focal plane, the reconstruction of the trajectory of one of the fragments (PRISMA arm) passing through well known magnetic fields (a dipole and a quadrupole) or the Bragg peak analysis (Bragg chamber arm).

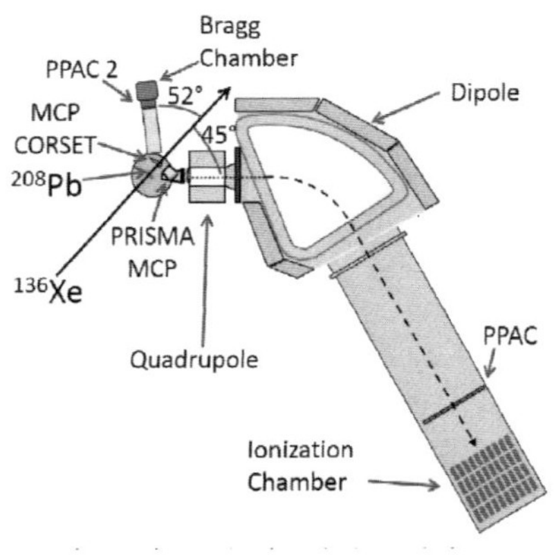

Figure 5. PRISMA at 45° and Bragg-arm at 52° with respect to the beam. The first consists of a MCP, quadrupole and dipole magnets, a PPAC in the focal plane and a $\Delta E - E$ telescope-like ionization chamber. The Bragg-arm has a MCP start and a PPAC stop, with a Bragg chamber behind the latter.

The preliminary analysis focused on the kinematic reconstruction, including the design and application of an algorithm of energy correction due to the losses in target and start detector which takes into account the change of charge state of the ions [6].

The observables $V_{\parallel}$ and $V_{\perp}$, projections of velocity vectors onto the reaction plane and on an orthogonal axis, are the key to search for full momentum transfer binary events, since for them $V_{\parallel} = V_{cm}$, $V_{\perp} = 0$. The possibile cases are shown in the Kinematic Diagram in Fig. 6:

(a) $V_{\parallel} = V_{cm}$, a FMT binary event for which the mass reconstruction is well interpreted;

(b) $V_{\parallel} > V_{cm}$, sequential fission (usually) of the TLF occurs and the secondary fragment emitted forward is detected, leading to a misinterpretation of the velocities;

(c) $V_{\parallel} < V_{cm}$, the secondary fragment emitted backward is detected.

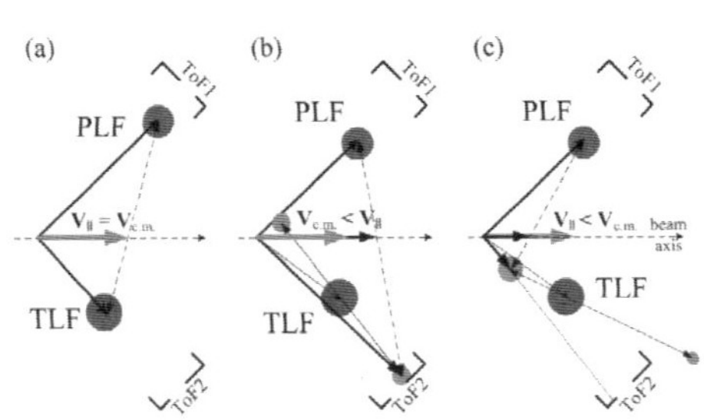

Figure 6. Kinematic Diagram [1]. (a) a FMT binary event; (b) sequential fission of TLF, the secondary fragment emitted forward is detected; (c) fragment emitted backward is detected.

In Fig. 7, the experimental $V_{\parallel}$ - $V_{\perp}$ distribution shows a peak corresponding to FMT binary events, which can be selected by application of a narrow gate around $V_{\parallel}/V_{cm} \sim 1$ and $V_{\perp} \sim 0$,

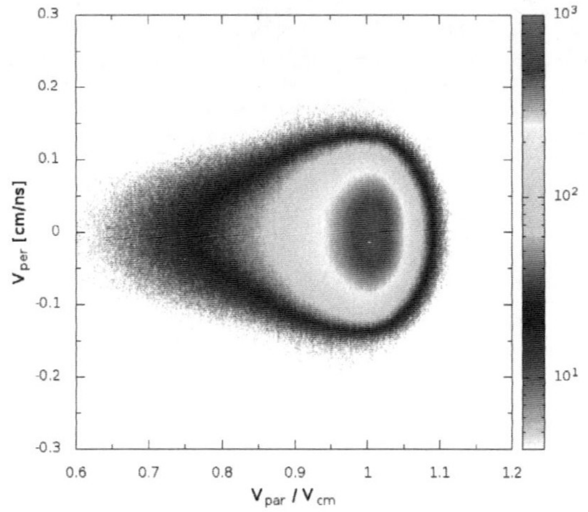

Figure 7. $V_{\parallel}$ - $V_{\perp}$ experimental distribution of ^{136}Xe+^{208}Pb reaction. A gate around $V_{\parallel}/V_{cm} \sim 1$ and $V_{\perp} \sim 0$, must be applied in order to discard sequential fission events.

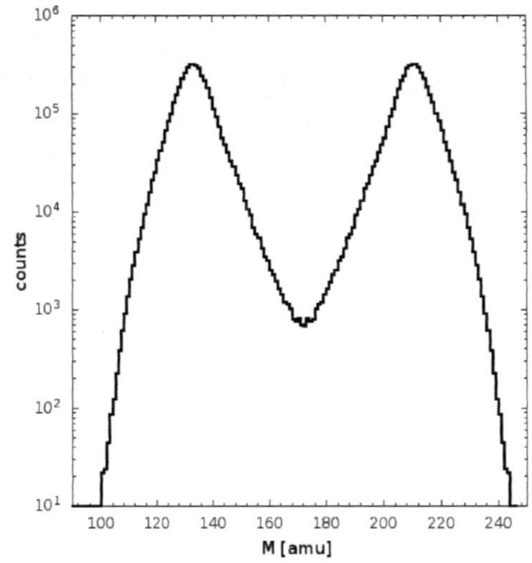

Figure 8. Preliminary results of ^{136}Xe+^{208}Pb reaction. The peaks are around the masses of Xe and Pb. There are cases of asymmetrizing transfers ending up in fragments heavier than target.

discarding sequential fission. The preliminary results of Fig. 8 show the mass of primary fragments obtained after the application of a gate on the $V_{\parallel}$ - $V_{\perp}$ matrix.

The mass distribution (Fig. 8) is double hampered. The two peaks are centered around the masses of projectile and target. Mass transfers from heavier partner to lighter one lead to a symmetric configuration of the output fragments and populate the masses between xenon and lead but it is clearly visible the possibility of transfers in the other way, increasing the mass asymmetry and leading to species heavier than the target.

6. Conclusions

The observed shell-effect persistence of the *magic numbers* N=82,126 makes the multinucleon transfer channel really interesting for the production in the *Terra Incognita* and, in this specific case, in the region with N=126, of great importance for a better understanding of the r-process.

When TLFs havier than Pb are produced, Q_{gg} is negative and lowers the excitation energy. This means that mass transfers from xenon projectiles to lead target end up in fragments with greater survival probability against neutron evaporation. Since masses way grater than 200 amu have been produced, this is an exciting result, showing how this reaction channel is to be chosen as a candidate in the attempt of producing heavy, *surviving*, neutron-rich nuclei, not reachable by other methods. Though the analysis is only at a preliminary stage, this is an important result and the experimental program is going to continue with reactions between heavier partners, like Xe+Cm and also implying lighter partners in order to search the same stabilizing effects in smaller closed shells such as N=28,82. The further step will be the application of techniques for the atomic number evaluation and, in the near future, the exploitation of radioactive ion beams.

Acknowledgements

One of the author (DQ) would like to thank the whole collaboration and in particular the spokepersons of the experiment E. Kozulin and L. Corradi.

IOP Conf. Series: Journal of Physics: Conf. Series **1014** (2018) 012015 doi:10.1088/1742-6596/1014/1/012015

References

[1] Kozulin E M, Vardaci E *et al* 2012 *Phys. Rev.* C **86** 044611
[2] Zagrebaev V and Greiner W 2008 *Phys. Rev. Lett.* **101** 122701
[3] Corradi L *et al* 2002 *Phys. Rev.* C **66** 024606
[4] Zagrebaev V and Greiner W 2011 *Phys. Rev.* C **83** 044618
[5] Zagrebaev V and Greiner W *J.Phys.* G **35** 125103
[6] Schiwietz G and Grande P L 2001 *Nucl. Instr. and Meth. in Phys. Res.* B **175-177** 125
[7] Bass R 1980 *Nuclear Reactions with Heavy Ions* (Berlin: Springer) p 326

IOP Conf. Series: Journal of Physics: Conf. Series **1014** (2018) 012016 doi:10.1088/1742-6596/1014/1/012016

Status and Perspectives of the INFN-LNS In-Flight Fragment Separator

P. Russotto[1], L. Calabretta[1], G. Cardella[2], G. Cosentino[1], E. De Filippo[2], B. Gnoffo[2,3], M. La Cognata[1], N.S. Martorana[1,3], E.V. Pagano[1,3], R.G. Pizzone[1], L. Quattrocchi[2], S. Romano[1], A.D. Russo[1] and D. Santonocito[1]

[1]INFN-Laboratori Nazionali del Sud, Catania, Italy
[2]INFN-Sezione di Catania, Catania, Italy
[3]Dipartimento di Fisica e Astronomia Università, degli Studi di Catania, Catania, Italy

E-mail: `russotto@lns.infn.it`

Abstract. In the last 15 years the FRIBs@LNS facility has successfully produced Radioactive Ion Beams using the In-Flight technique. We report on the current status and future perspectives opened by FRAISE, a new fragment separator that will be build in connection with the upgrade of Superconducting Cyclotron of the INFN-LNS laboratories.

1. Introduction

In the last decades several facilities devoted to the production of Radioactive Ion Beams (RIBs) for experimental studies in nuclear physics have been developed worldwide [1–5] and many other are still under construction [6–9]. These facilities have been based on two main production techniques: the Isotope Separation On Line (ISOL) [10] and the In-Flight [11] production methods. Thanks to progress in producing RIBs with these methods, experiments employing short-lived isotopes have become feasible and widely used; therefore, many properties of unstable nuclei have been investigated [12–14]. However, there is still room for studies in this field, since the full perspectives opened by the use of RIBs are far from being fully exploited. In particular, coupling progress in production methods with advances in experimental techniques and detection devices can still open new interesting perspectives, allowing to carry out better or new experiments.
At the INFN-LNS [15] in Catania (Italy) a facility using the in-flight technique to produce RIBs has been operating for the last 15 years. Important results have been achieved as briefly discussed in the next section. The project of an upgrade of the LNS-INFN Superconducting Cyclotron, leading to an increase of the power of available primary beams, will open new perspectives for the production of RIBs. This will require the installation of a new Fragment Separator, whose main structure and principle of operation are briefly described in section 3. A variety of physics cases could then be investigated in the future, thanks to the performances of upgraded fragment separator and the detection devices already existing or in construction at the INFN-LNS. Few examples will be discussed in section 4.

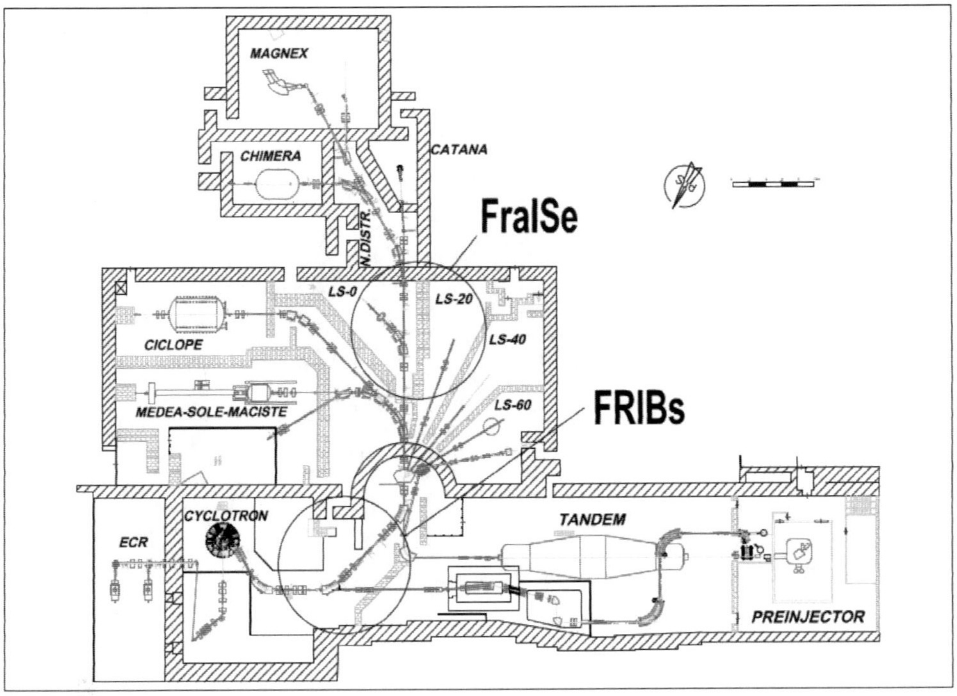

Figure 1. Schematic view of the INFN-LNS beam lines and halls. The position of the actual fragment separator, FRIBs, and of the new one, FraISe, are indicated by circles.

2. The FRIBs facility

The FRIBs@LNS (in Flight Radioactive Ion BeamS at LNS) facility [16], presently installed at the INFN-Laboratori Nazionali del Sud, allows one to carry out nuclear physics experiments investigating the properties of short-lived nuclear species. The RIBs are produced by using the so called in-flight technique, that is, the fragmentation of a fast projectile (the primary beam accelerated by the LNS Superconducting Cyclotron) on a usually thick Beryllium production target (placed near to the exit of the Superconducting Cyclotron). Since a multitude of (mainly stable) nuclei are produced in such a process, a magnetic analysis/selection is needed to single out RIBs of interest for the specific study. In FRIBs we use just two 45° dipoles of the standard transport line to perform the magnetic selection, trying to maximize the yield of the wanted unstable isotope and eliminate unwanted stable isotopes and the primary beam. In such a way, since 2001, following the pioneering work of G. Raciti and his collaborators, rare isotope from ^{6}He to ^{68}Ni have been produced and effectively used in several experiments. Produced RIBs were (or could be) delivered to many experimental halls, housing dedicated detectors like CHIMERA [17] and FARCOS [18], Magnex [19], Hodo-Ct [20] or to other multi-purpose scattering chambers. A schematic view of the INFN-LNS beam lines and halls is shown in Fig. 1; FRIBs is indicated by a circle.

Since a magnetic analysis cannot separate isotopes having the same m/q ratio (m/Z in case of fully stripped ions) and has always a finite dp/p acceptance ($\sim 1\%$ in the case of FRIBs) we obtain a cocktail beam, i.e. it contains several isotopes at the same time. In some cases, this can also be an advantage since a single experimental configuration allows to carry out a measurement with different RIBs at the same time. However, an event by event identification (tagging) of the ion arriving on the target point is needed in order to off-line select the wanted projectile.

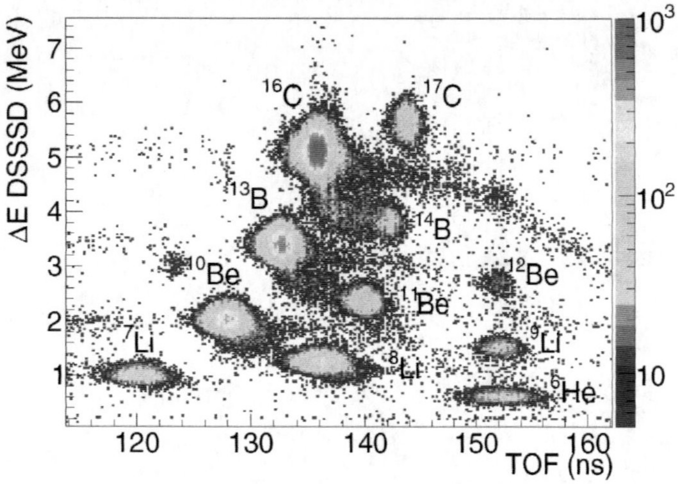

Figure 2. $\Delta E - ToF$ correlation plot, as given by the tagging system described in the text, for the cocktail beam produced by ^{18}O beam of 55 $AMeV$ on a 1500 μm thick ^9Be target. Extracted from Ref. [21].

For that, an advanced device [22] has been used in the CHIMERA beam line; it consists of a large-surface Micro Channel Plate detector placed about 13 meters before the entrance of the scattering chamber and producing the start of Time of Flight (ToF) measurement, and a thin (140 or 300 μm) Double Sided Silicon Strips Detector, with 32 (64×2 mm^2) strips on both horizontal and vertical planes, placed just before the entrance of the scattering chamber, delivering energy loss (ΔE), position and stop of ToF measurements. By combining the ΔE and ToF information, an event-by-event identification of arriving ion is then possible. As an example, the ΔE-ToF correlation plot for the cocktail beam obtained through the fragmentation of a primary ^{18}O beam of 55 $AMeV$ energy on a 1.5 mm thick ^9Be target is shown in Fig. 2. This was obtained with a magnetic rigidity selection of $B\rho \sim 2.8$ Tm. Since the magnetic rigidity is almost fixed, ions having the same m/q ratio have the same velocity and stay along a vertical line (constant ToF) in such a plot. In addition, also a Parallel Plate Avalanche Counter position sensitive detector, placed between the Silicon Strips Detector and the target station, has been used to get a trajectory measurement.
Typical intensities available with FRIBs, for nuclei not very far from stability valley (e.g. $^{10}Be, ^{16}C, ^{18}Ne, ^{68}Ni$) have been of about $10^{4-5}pps$; these intensities have been mainly limited by the maximum power (100 W) achievable with the actual Superconducting Cyclotron properties. Nevertheless, important physics cases have been investigated by using the RIBs delivered by FRIBs. As few examples, here we want to remind some important results. One was the discovery of an important diproton radioactivity contribution from ^{18}Ne excited states by G. Raciti and his collaborators [20]. It is worth mentioning also the investigation of the cluster structure of ^{10}Be by D. Dell'Aquila et al. [21], where, measuring the $\alpha + ^6He$ break-up channel, the ^{10}Be excitation energy spectrum was reconstructed and the existence of a new 6^+ level around 13.5 MeV of excitation energy was found. Recently, also the Pigmy Dipole Resonance in radioactive neutron rich ^{68}Ni nuclei, excited by means of an iso-scalar ^{12}C probe, has been carefully studied [23], as reported in another contribution to this conference [24].
However, due to limited maximum power delivered by the Superconducting Cyclotron, investigations have always been limited to nuclei not far away from stability valley, being the

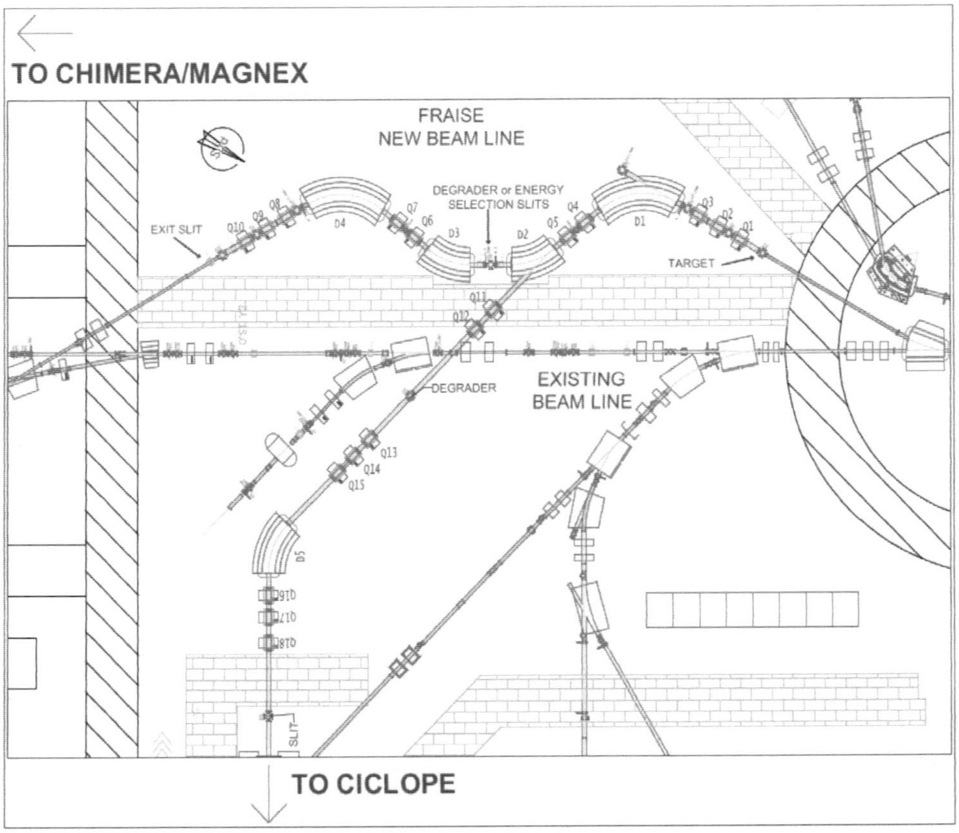

Figure 3. Schematic view of FRAISE the new fragment separator of the LNS.

yields for very exotic nuclei not sufficient for a realistic "time-limited" experiment.

3. FRAISE: the new FRAgment In-fligth SEparator of INFN-LNS

In the forthcoming years, the Superconducting Cyclotron of the INFN-LNS is going to be renewed and improved [25]. One of the main novelties will be the replacement of the actual extraction system, based on electrostatic deflector, with a new extraction system based on stripping. Extraction by stripping is based on the sudden change of the magnetic rigidity of the accelerated ion, when its charge state is suddenly increased crossing a thin carbon foil. This method will be highly efficient for light ions; indeed, ions with mass number $A < 20$ and energy higher than 15 $AMeV$ are fully stripped, q=Z, with probability $> 99\%$ when crossing a stripper foil with equilibrium thickness [26]. However, the new extraction method will be very efficient also for medium mass nuclei up to $A \sim 40$. Intensities of the order of 10^{13} pps will be reached for beams, from C to Ar, at energies of $30 - 70$ $AMeV$; typical beam power will be of ~ 5 kW. This upgrade project of the Superconducting Cyclotron opens new perspectives also for the production of fragmentation beams. The expected availability of higher power for primary beams can be used to produce very intense RIBs, allowing to obtain good intensities also for ions very far from stability valley. Due to the high intensity beams delivered by the upgraded Cyclotron, radio-protection issues impose to place the fragment separator in a position different from the present one, i.e. the accelerator vault, as shown in Fig. 1; in fact it is not possible to efficiently shield this area [27]. Therefore, we have started to study the design of a new fragment

separator, named FRAISE (FRAgment In-flight SEparator), explicitly designed to profit of this intensity upgrade. Since the idea is to maintain the possibility to deliver the beam in the largest possible number of target positions, we decided to place the new fragment separator in the area in which are currently located the 20° and 40° LNS experimental halls; this area is shown as a circle in Fig. 1. According to performed simulations, up to 2 kW of primary beam power can be efficiently and safely used in this new configuration [27]. Detailed information about technical features and expected performance of FRAISE are available in Refs. [28].
A new production target, able to dissipate the heat absorbed, will be designed. After that, the first part of the fragment separator will be made up of two 70° and 40° dipoles (D1 and D2, respectively, in Fig. 3) and 1 triplet and 1 doublet (Q1-Q3 and Q4-Q5, respectively, in Fig. 3). The second part of the fragment separator will just be a mirror-copy of the first part. This is schematically shown in Fig. 3. This design ensures a full achromaticity of the fragment separator.
However, on the symmetry-dispersive plane we plan to have a mechanical control of beam profile by means of a mechanichal slits system, and the possibility to insert, just after the slits system, a wedge (degrader) for performing a better rejection of unwanted ions. In fact, a magnetic spectrometer cannot separate ions having the same A/q ratio (A/Z in case of fully stripped ions); one solution is then to use a degrader, i.e. a piece of matter inserted at the dispersive focal plane between the two sections of the fragment separator. Because the atomic slowing down of the ions is roughly proportional to Z^2/v^2, different isotopes, depending mainly on their atomic number, will have different velocities after passing through the degrader. This change in velocity causes the isotopes with the same A/Z ratio to be separated in the second dispersive stage. Detailed calculations [29] show that this second selections is roughly sensitive to A^3/Z^2. This method has been described elsewhere [1] and requires that the material is shaped as a wedge or with thickness increasing properly in order to preserve the achromaticity of the separator. This can be simply understood qualitatively since for a given fragment different positions in the dispersive focal plane correspond to different velocities, and then the energy loss must be adjusted accordingly by varying the thickness: a thicker degrader is needed at the higher velocity side and a thinner one at the lower velocities. A degrader that is shaped accordingly is called an "achromatic" degrader. Because the degrader slows all particles down, including the selected isotope, the $B\rho$ setting of the magnetic elements following the degrader have to be accordingly adjusted. In particular, when a degrader is used, the optical settings of the two dispersive sections of the fragment separator are not mirror of each other anymore, and the magnetic rigidity of the second dipole has to be decreased. In other cases, it can be desirable that the shape of the wedge preserves another parameter of the beam such as, for instance, its velocity. In that case, the wedge is called "monochromatic" and is shaped in order to narrow the energy spread of the selected particles. This feature can be used in experiments where the incident energy of the RIBs must be well known as, for example, in spectroscopy experiments. It should be noted that using a wedge always reduces the rate of the produced RIBs. At the exit of the fragment separator, a second slits system will allow a further selection/control of the beam profile. This section of the FRAISE will be then able to deliver RIBs toward the CHIMERA and MAGNEX halls.
Since we want to maintain the possibility to deliver the RIBs in other target positions, we plan to build also a second arm of FRAISE towards the already existing CICLOPE large area multi-purpose scattering chamber. This solution is achieved operating the second dipole D2 to bent the beam at $-5°$ instead of the 40° of the previous configuration; an additional 45° (D5) dipole will drive the beam towards CICLOPE. Also in this case we foresee to have a slits system on the intermediate plane, followed by a wedge and a second slits system at the exit. Also one doublet (Q11-Q12) and two triplets (Q13-Q15 and Q16-Q18) need to be added. The possibility of placing the SOLE superconducting solenoid [30] before the CICLOPE chamber has been taken

IOP Conf. Series: Journal of Physics: Conf. Series **1014** (2018) 012016 doi:10.1088/1742-6596/1014/1/012016

into account; SOLE could act as a beam focusing element on the CICLOPE target station, but could also be used as a stand-alone device operated in a configuration similar to the one of the HELIOS spectrometer at Argonne [31, 32], as discussed below.
Also a revision and improvement of the diagnostic systems [33, 34] will be needed to measure the intensity and the profile of the RIBs cocktail along the lines, in order to perform an optimal transport towards the target stations.
This new design is expected to be able to transport a larger fraction of produced beams, having a larger and selectable momentum acceptance up to a $\Delta p/p \sim 4\%$, to be compared with the old solution having a fixed $\Delta p/p \sim 1\%$. With an efficient rejection of the unwanted species from the cocktail beam, by proper using the slits systems and the wedge, RIBs produced will have an higher purity. In many cases, it will be possible to get higher intensities for the wanted isotope, with respect to FRIBs, still keeping the total rate low enough to allow the use of the present tagging system. Expected intensities, calculated by using the LISE simulation code [35], are listed in table 1.
In order to efficiently exploit the projectile fragmentation mechanism primary beams of 40-60 AMeV of incident energy are commonly used; it follows that RIBs at 20-45 AMeV (after the wedge and tagging steps) will be produced. However, as shown in that table, we are also taking into accounts the possibility to use very thick tagging detectors or additional degrader foils, to get RIBs at few AMeV; this energy regime is interesting for nuclear astrophysics studies, as discussed below; in these cases, a proper refocusing of the RIBs after the degrading could also be required.

Table 1. Expected yield with the new fragment separator FRAISE

main product	primary beam/energy (AMeV)	thickness Be target (μm)	wedge thickness (μm)	primary intensity (kW)	expected yield (kHz)	purity (%)	energy after tagging (AMeV)
^{14}Be	$^{18}O/55$	1500	0	2	2.6	2	46
^{14}Be	$^{18}O/55$	1500	1000	2	2.2	70	43
^{13}N	$^{16}O/40$	700	600	2	1230	54	4
^{14}O	$^{16}O/40$	700	600	2	807	36	4
^{18}Ne	$^{20}Ne/60$	1000	0	2	16700	16	43
^{18}Ne	$^{20}Ne/60$	1000	1000	2	3120	47	24
^{17}F	$^{20}Ne/60$	1000	1000	2	3300	49	23
^{34}Si	$^{36}S/40$	500	500	2	980	81	11
^{38}S	$^{40}Ar/40$	500	300	2	1840	66	17
^{34}Ar	$^{36}Ar/50$	250	0	2	2800	4	41
^{34}Ar	$^{36}Ar/50$	250	500	2	426	12	4
^{68}Ni	$^{70}Zn/50$	250	200	1	490	50	18

4. Some FRAISE physics cases

4.1. Isospin physics

The study of Isospin effects in Heavy-Ion reactions at Fermi energies has been an important field of research of LNS laboratory in the last 15 years [17, 36]. As a relevant example, thanks to the

IOP Conf. Series: Journal of Physics: Conf. Series **1014** (2018) 012016 doi:10.1088/1742-6596/1014/1/012016

capabilities of the CHIMERA multi-detectors [17] careful studies of the reaction mechanisms have been pursued.
In the Limiting experiment, the competition between incomplete fusion and binary reactions in central collisions of medium mass nuclei (^{40}Ca+^{40}Ca,^{46}Ti,^{48}Ca and ^{48}Ca+^{48}Ca) at 25 $AMeV$ was investigated [37–39]. It resulted that the incomplete fusion was more probable, against the binary reactions, for the neutron rich-systems. By using the CoMD-II dynamical model [40], it was demonstrated that the observed effect depends on the stiffness of the symmetry energy and, by comparing the obtained mass spectra with prediction of the model, an effective constrain for the symmetry energy was obtained. These systems were also used for a careful study of the Isospin equilibration phenomena [38, 39], studying the isobaric $^7Li/^7Be$ and $^3H/^3He$ ratio); by analyzing the crossed systems (^{48}Ca +^{40} Ca and viceversa) it was there found that complete equilibration does not take place at these energies. The ^{40}Ca+^{40}Ca ($N/Z = 1$) and ^{48}Ca+^{48}Ca ($N/Z = 1.4$) represent a couple of systems with very large Isospin difference, an ideal case for investigating Isospin and symmetry energy effects. However, the new FRAISE will be able to offer better cases. In fact, in this mass region, neutron rich-poor $^{46,34}Ar$ of about 10^6 pps will be produced by using primary ^{48}Ca and ^{36}Ar beams, respectively. This will allow to further extend the Isospin differences ($N/Z = 1.56$ for ^{34}Ar and 0.89 for ^{34}Ar) and to study above mentioned effects in very exotic conditions.
In the region of heavier projectile-target, the Intermediate Mass Fragments production was carefully studied in ^{124}Sn+^{64}Ni, ^{112}Sn+^{58}Ni, ^{124}Xe+$^{64}Ni(Zn)$ at 35 AMeV semi-peripheral collisions [41, 42]. There it was shed light on the mechanisms of dynamical fragment production, proposing the scenario of the fast fragmentation-rupture of the neck, connecting projectile and target like fragments at the beginning of their re-separation, as mechanism of lighter IMF emission. Other dynamical mechanisms were evidenced in the break-up of the excited PLF*, as evidenced by the observation of non-isotropic angular distributions for the emission direction of break-up products [41]. By looking at systems with different Isospin content, an higher probability of occurrence of the dynamical emission was found for the neutron rich systems, thus envisaging a role of the Isospin content on triggering the conditions leading to the dynamical emission mechanisms. These reactions have also been used to investigate the density dependence of the symmetry energy at densities smaller than the saturation one, by studying the neutron enrichment of the neck emitted fragments [43].
In these cases, the use of very neutron rich projectile as the ^{68}Ni, already produced with the old FRIBs, will allow to reach higher Isospin content for the entrance reaction channel. Moreover, it is now possible to greatly enhance the energy and angular resolutions of such studies by coupling CHIMERA with the new FARCOS [18] device.
It follows that the FRAISE RIBs, in these regions of medium mass nuclei, will represent an important opportunity for extending studies, already performed at LNS, toward high Isospin asymmetries. We remind here that in these studies, symmetry energy effects depends on the square of the Isospin asymmetry; higher Isospin asymmetries allows then to enhance the effects and to separate these from those related to the properties of symmetric matter. This is of some importance in order to get observation that, through comparison with transport model [40, 44], can shed light on the properties of the effective in-medium nucleon nucleon interaction.

4.2. Cluster physics
In the recent years the study of cluster structures of nuclei has become one of the main topic of nuclear physics [45]. Cluster structures have been well evidenced in stable light nuclei, especially for the alpha-conjugate nuclei as well described in the Ikeda theory-diagram [46]. One important issue is to study how the cluster structure evolves when valence particles (either protons or neutrons) are added [47]. In this instance there exists the possibility that the valence particles - most typically neutrons - may be exchanged between the alpha-particle cores [47]. This type

of exchange process is completely analogous to the exchange of electrons in atomic molecules. This leads to the concept of nuclear molecules, with alpha particles playing the role of core structure and neutrons acting as covalent bonding between them. This is well described in the modified Ikeda theory-diagram [46]. In this framework, nuclei arranged as nuclear molecules are expected to have very deformed configurations, characterized by high angular momentum and strong rotational band structure in the excitation energy spectrum. This has been already studied in several Be isotope [48], including the case of ^{10}Be [21] discussed in sect. 2. Another interesting case is the one of C isotopes; possibly the best case for the linear arrangement of cluster, from a theoretical perspective, is ^{16}C [49], though in this instance experimental data is very sparse. As an example the break-up of ^{16}C in ^{6}He+^{6}He+^{4}He has been already investigate with FRIBs, in the same experiments studying the ^{10}Be case [21]. The collected statistic was enough to extract the ^{16}C excitation energy spectrum and to speculate on new levels, but still not enough for a quantitative assignment. With FRAISE the yield of ^{16}C will increase by a factor $\sim$ 20; this kind of measurements will become then easily feasible. Many others unstable light nuclei, expected to have molecular or halo structure, will be produced and made available for effective experiments. Also in this case, the use of FARCOS array coupled with CHIMERA multi-detector will allow to get very high energy and angular resolutions, high efficiency and excellent identification capabilities.

4.3. Nuclear Structure

Nuclear structure studies in nuclei far from stability valley usually requires both high efficiency detector array and high cross sections reactions. Direct reactions represent the tool of choice as they allow to infer basic properties of investigated nuclei: elastic scattering can probe protons and neutrons density distributions, inelastic scattering can probe collective aspects, and transfer reactions allow to get information on single particle states and couple-correlation. The necessity to use inverse kinematic, in the case of RIBs, introduces some experimental problems both in the identification of low energy particles emitted in such kind of reactions and in the angular resolution, since the strong focusing of products in a limited angular region tends to reduce the achievable separation among the excited states. An approach allowing to overcome such difficulties is to use an array of segmented detectors placed in a homogeneous magnetic field generated by a solenoid, following the scheme already used by the Helios spectrometer at Argonne [31, 32]. The basic idea is to place the target inside the solenoid along the magnetic axis, and focus the charged particles from the two-body reaction on an array of silicon detectors placed along the same magnetic axis of the solenoid. Measuring energy, time of flight and the impact point, it is then possible to identify the emitted particle and reconstruct the emission angle. In fact the Helios spectrometer has used this new approach for the measurement of light charged particle and has been successfully used in several experiments, proving the validity of this method and the advantages in terms of resolution and detection efficiencies. An R&D study on possibility to build a Helios-like spectrometer at the LNS, using the already existing SOLE solenoid [30], presently placed after the MEDEA detector [50], has already been started. Such kind of spectrometer represents an ideal instruments for measuring direct reactions induced by the RIBS produced with FRAISE. As a relevant example, we just discuss briefly the case study of elastic and inelastic scattering of ^{8}B on protons. ^{8}B is a proton rich nucleus, candidate for being a proton-halo nucleus with a proton separation energy of 137 keV. Experimental evidences of the proton halo structure are not straightforward [51–54] and comparison of experimental data and simulations leads to controversial conclusion. The ^{8}B isotope can be produced using a primary ^{12}C beam at 55 $AMeV$ on a thick 1500 μm ^{9}Be target. Assuming a primary beam intensity of 2 kW, a 1000 μm thick Al wedge, and a selective slits setting of $\Delta p/p \sim 2\%$, we expect to be able to deliver $1.9 \times 10^{5} pps$ of ^{8}B, over a total cocktail beam rate of 3.3×10^{5}, with a purity of $\sim$ 60% with only ^{7}Be as strong contaminant. The expected $\Delta E - ToF$ correlation, using a tagging

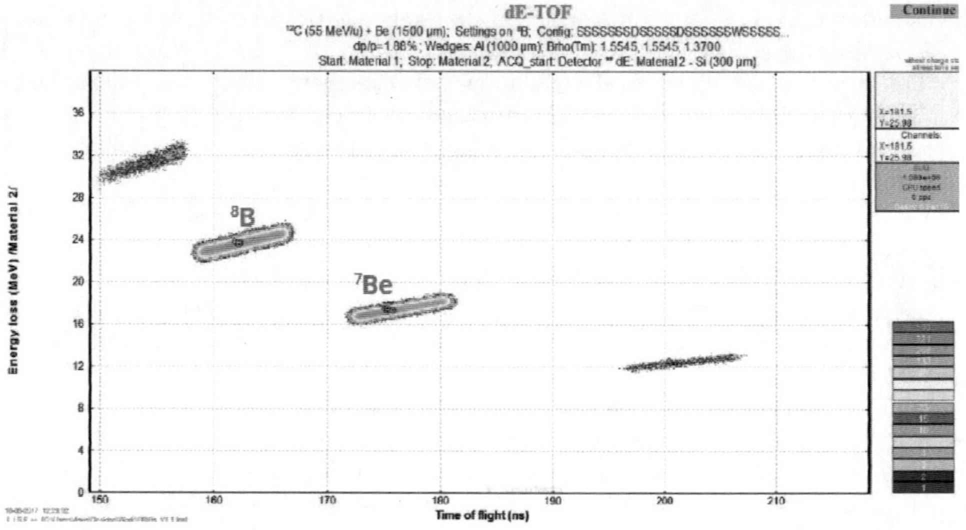

Figure 4. expected $\Delta E - TOF$ tagging of the cocktail beams produced by ^{12}C of 55 $AMeV$ energy on ^{9}Be target.

system similar to the one now in operation, is shown in Fig. 4. A clear separation between the species is obtained. This will make feasible a (p,p') scattering measurement. Moreover, by using a Helios-like spectrometer it would be possible to reach, with respect to previous measurement, a higher efficiency for measuring the angular distributions and a better resolution of the excited states.

Another interesting case is the study of the scattering of ^{11}Be on protons. ^{11}Be has received particular interest because of the inversion between $p_{1/2}$ and $s_{1/2}$ orbitals [55] and has a neutron-halo structure, with resonances above the particle emission threshold. Also in this case we expect to reach about $1.5 \times 10^5 pps$ with FRAISE. Studies of (p,p') elastic and inelastic scattering at MSU allowed to populate excited states up to about 8 MeV [56]. However, due to the very small separation between ground and first excited state (320 keV), energy level have not been separated. Moreover, the excitation energy spectrum shows several not resolved resonances and, up to now, the angular distributions have been studied for energy bins containing several levels. Comparison with theoretical calculations do not reproduce the inelastic angular distributions [56]; a possible cause could be connected to the ^{10}Be core excitation contribution. A Helios-like spectrometer would make the separation of these excited states possible allowing for a deeper understanding of the ^{11}Be structure.

4.4. Nuclear Astrophysics

With FRAISE we foresee also to extend the use of RIBs towards the low energy domain, in order to exploit the possibility of indirect measurements having astrophysical interest. In fact, many interesting projectiles could be delivered, allowing to estimate reaction rates of great interest for explosive nucleo−synthesis calculations. Explosive nucleo−synthesis takes place in some of the most spectacular and peculiar scenarios of the cosmos. The main feature of such phenomena, from the nuclear astrophysics point of view, is the fact that the nucleo−synthesis takes place through the realm of exotic nuclides at energies which are sensitively below the Coulomb barrier. In such environments, in fact, the average time for a reaction to take place is much shorter than the decay time of the involved isotopes. Among such explosive scenarios,

Type I X-ray bursts [57–59,62] are observed as sudden, intense emissions of X-rays. These bursts are very interesting phenomena especially because they might turn out to be also very useful probes of our cosmos [58, 59] and a very useful tool to constrain neutron star models. With this respect, we want to highlight the importance of the studies in those fields mentioning the recent discoveries of a binary neutron–star coalescence through gravitational waves [60] and of heavy elements production in the same sites [61]. Of course, this might be possible only if an exhaustive understanding of the development of the burst is achieved. But, up to now, most of the reaction rates adopted in nucleo–synthesis calculations rely on theoretical estimates using statistical models and, therefore, may be subject to significant uncertainties since levels densities are not large enough to allow for their application .
A relevant example is the $^{14}O(\alpha,p)^{17}F$: this reaction could be the onset of a possible route that breaks out from the HCNO cycles, leading to the build-up of new elements. It follows that the knowledge of this reaction cross section is of fundamental importance to estimate the rate of heavy element production. But, even with a high rate RIBs, cross section at the relevant energies, between 1-2 MeV in the center of mass, will be hard to measure. In fact, such energies are quite smaller than the Coulomb barrier (4.5 MeV, in the case of the $^{14}O(\alpha,p)^{17}F$): since nuclear reactions take place through tunneling of this barrier, cross sections are very small and the signal-to-noise ratio approaches zero. This is even more critical in the case of RIBs, as beam intensities hardly exceed 10^6 pps, several orders of magnitude less than typical stable beams. It follows that cross sections at astrophysical energies are extrapolated from higher energies where data are available. However, systematic uncertainties due to extrapolation might jeopardize model predictions. Moreover, in the case of direct measurements, the ejected proton (or/and ^{17}F) has to be detected: this has to face many problems (small counting rate, low statistics, large background, energy and angular resolution).
Indirect methods, such as the Trojan Horse Method (THM), developed by the LNS nuclear astrophysics group [63, 64], have then been devised to bypass these shortcomings. The basic idea underlying indirect methods is the use of nuclear reaction theory to link the cross section of astrophysical importance to the one of a different process, easier to study with present-day facilities. The THM has proved its effectiveness in the extraction of cross sections of reactions having charged particles or neutrons in the exit channel, allowing to achieve astrophysical energies with no need of extrapolation. In the THM approach, the low-energy cross section of a $A(x,b)B$ reaction is obtained by extracting the so called quasi-free contribution to a suitable $A(a,bB)s$ reaction. Therefore, a careful analysis of the reaction dynamics and of the role of clusters is necessary before applying the method. The application of the recent developments of the THM formalism to reactions induced by RIBs would open a new frontier in nuclear astrophysics and in nuclear structure studies. The method itself has been already successfully applied to exotic beams [65–67] and the possibility of having low-energy exotic beams at LNS could give a unique opportunity for the nuclear astrophysics community. Targets containing ^6Li such as ^6LiH or ^6LiF can be used as a source of virtual α-particles, coming from ^6Li direct break-up. A great advantage of this project is the possibility to use beam energies larger than the ones needed for direct measurements, as prescribed by THM. Silicon strip detector array could be used to detect the emitted particles, guaranteeing excellent energy and position resolution. The region relevant for astrophysics, located below about 1.5 MeV, can be fully covered as well as the region at about 2 MeV where direct data are often available for comparison and method validation. Besides particle identification, e.g., by means of the standard DE-E technique, the channel of interest can be singled out by the coincidence measurement of the outgoing particles and the extraction of the reaction Q-value spectrum.
Other interesting reactions, that could be studied with similar methods and experimental apparata are the $^{18}Ne(\alpha,p)^{21}Na$, another important route toward heavy elements nucleo–synthesis, the $^{34}Ar(\alpha,p)^{37}K$ reaction, since ^{34}Ar is a waiting point in X-ray bursts

nucleo–synthesis [68, 69] and the $^{13}N(\alpha,p)^{16}O$ reaction, since in asymptotic giant branch stars this reaction might be an additional n-source for heavy element production through slow neutron capture (s-process); the ^{13}C supply might be reduced if ^{13}N is burnt to ^{16}O before it decays thus influencing the s-process.
Thus, the possibility to measure RIBs induced reaction cross sections at low energies with the THM will represent a step forward in the physics programs of the INFN-LNS.

5. Acknowledgments

The authors wish to thank the NewCHIM collaboration, for the continuous work and support to the FRIBs facility, and the ASFIN2 and LNSSTREAM2 collaborations for their contribution to the discussion of physics cases for the new Fragment Separator.

References

[1] Anne R *et al* 1987 *Nucl. Instr. Meth.* A **257** 215
[2] Geissel H *et al* 1992 *Nucl. Instr. Meth.* B **70** 286
[3] Kugler E *et al* 1992 *Nucl. Instr. Meth.* B **70** 41
[4] Kubo T *et al* 1992 *Nucl. Instr. Meth.* B **70** 309
[5] GANIL/SPIRAL 2001-2007 Achievemnets, Highlights and Perspectives http://hal.in2p3.fr/in2p3-00336915/
[6] see http://www.fair-center.eu/
[7] see http://frib.msu.edu/about/index.html
[8] Prete G 2010 The LNL radioactive beam facility *Scholarpedia* **5(5)** 9751
[9] see http://pro.ganil-spiral2.eu/spiral2
[10] Van Duppen P 2006 Isotope Separation On Line and Post Acceleration *Lect. Notes Phys.* **700** 37-77
[11] Blumenfeld Y, Nilsson T and Van Duppen P 2013 *Phys. Scr,* **T152** 014023
[12] Tanihata I 1995 *Prog. Part. Nucl. Phys* 33 505-573
[13] Casten R F and Sherrill B M 2000 *Prog. Part. Nucl. Phys* **45** 171-233
[14] Hagino K 2012 Exotic nuclei far from the stability line *Preprint* http://arxiv.org/abs/1208.1583v2
[15] www.lns.infn.it
[16] Raciti G *et al* 2008 *Nucl. Instr. and Meth.* B **266** 4632
[17] Pagano A 2012 *Nucl. Phys. News* **22** 28 and refs. therein
[18] Pagano E V *et al* 2016 *Eur. Phys. J. Web of Conf.* **117** 10008
[19] Cappuzzello F, Agodi C, Carbone D and Cavallaro M 2016 *Eur. Phys. J.* A **52** 167
[20] Raciti G *et al* 2008 *Phys. Rev. Lett.* **100** 192503
[21] Dell'Aquila D *et al* 2016 *Phys. Rev.* C **93** 024611
[22] Lombardo I *et al* 2011 *Nucl. Instr. and Meth. (Proc. Suppl.)* B **215** 272
[23] Martorana N S *et al* 2018 *submitted to Phys. Lett.* B
[24] Martorana N S *et al* 2017 *see contribution to this workshop*
[25] Calabretta L *et al* 2017 *Mod. Phys. Lett.* A **32** 1740009
[26] Kunihiro S *et al* 1992 *Atomic Data and Nucl. Data Tables* **51** 173-241
[27] Russo S. (INFN-LNS) 2017 private communications
[28] Russo A D, Calabretta L, Cardella G and Russotto P 2017 *Preliminary design of the new FRagment Ions SEparator at INFN-LNS* contr. at *40th European Cyclotron Progress Meeting*, Sept. 2017 INFN-LNL Legnaro (Italy); Russo A D *et al* 2018 *LNS Activity Report 2015-2016* in press
[29] Dufour J P *et al* 1986 *Nucl. Instr. and Meth.* A **248** 267
[30] Bellia G *et al* 1996 *IEEE Trans. on Nucl. Sc.* **43** 1737
[31] Wuosmaa AH *et al* 2004 *Nucl. Phys.* A **746** 267c
[32] Wuosmaa A H *et al* 2007 *Nucl. Instr. and Meth.* A **580** 1290
[33] Amato A *et al* 2009 *LNS Activity Report 2009* 160
[34] Pagano E V *et al* 2018 *LNS Activity Report 2015-2016* in press
[35] Tarasov O B and Bazin D 2016 *Nucl. Instr. and Meth.* B **376** 185
[36] De Filippo E and Pagano A 2014 *Eur. Phys. J* A **50** 32 and refs therein
[37] Amorini F *et al* 2009 *Phys. Rev. Lett.* **102** 112701
[38] Cardella G 2012 *et al Phys. Rev.* C **85** 064609
[39] Lombardo I *et al* 2012 Jour. of Phys.: Conf. Ser. **381** 012093; Lombardo I *et al* 2013 Jour. of Phys.: Conf. Ser. **420** 012094.
[40] Papa M *et al* 2007 *Phys. Rev.* C **75** 054616 and refs therein
[41] De Filippo E *et al* 2016 *Il Nuovo Cimento* C **39** 404

IOP Conf. Series: Journal of Physics: Conf. Series **1014** (2018) 012016 doi:10.1088/1742-6596/1014/1/012016

[42] Russotto *et al* 2015 *Phys. Rev* C **91** 014610
[43] De Filippo E *et al* 2012 *Phys. Rev.* C **86** 014610
[44] Baran V, Colonna M, Greco V and Di Toro M 2005 *Phys. Rep.* **335** 410
[45] Freer M 2007 *Reports on Progress in Physics* **70** 2149
[46] Ikeda K, Tagikawa N and Horiuchi H 1968 *Prog. Theo. Phys. Suppl. extra number* **464**
[47] Freer M 2010 Clusters in nuclei *Scholarpedia* **5(6)** 9652
[48] von Oertzen W 1996 *Z. Phys.* A **354** 37.
[49] Itagaki N *et al* 2001 *Phys. Rev.* C **64** 014301
[50] Migneco E *et al* 1992 *Nucl. Instr. and Meth.* A **314** 31
[51] Minamisono T *et al* 1992 *Phys. Rev Lett.* **69** 2058
[52] Warner R E *et al* 1995 *Phys. Rev* C **52** R1166
[53] Schwab W *et al* 1995 *Z. Phys.* A **350** 283
[54] Kelley J H *et al* 1996 *Phys. Rev. Lett.* **77** 5020
[55] Talmi I and Unna I 1960 *Phys Rev. Lett* **4** 469
[56] Shrivastava A *et al* 2004 *Phys. Lett.* B **596** 54
[57] Galloway D K *et al* 2008 *Astrophys. J. Suppl.* **179** 360
[58] Galloway D K *et al* 2003 *Astrophys. J.* **590** 999
[59] Galloway D K *et al* 2006 *Astrophys. J.* **639** 1039
[60] Abbott B P *et al* 2017 *Phys. Rev. Lett.* 119 **161101**
[61] Abbott B P *et al* 2017 *Astrophy. J. Lett.* 848 **2**
[62] Strohmayer T E and Bildsten L 2006 *Compact Stellar X-Ray Sources* (Cambridge, UK: Cambridge Univ. Press)
[63] Spitaleri C *et al* 2016 *Eur. Phys. J.* A **52** 77
[64] Tribble R E *et al* 2014 *Rep. Prog. Phys.* **77** 106901
[65] Cherubini S *et al* 2015 *Phys. Rev.* C **92** 015805
[66] Pizzone R G *et al* 2016 *Eur. Phys. J.* A **52** 24
[67] La Cognata M *et al* 2017 *Astrophys. J.* **846** 65
[68] Parikh A *et al* 2008 *Astrophys. J. Suppl.* **178** 110
[69] Iliadis C 2007 *Nuclear Physics of Stars* (Wenheim, Germany: Wiley-VCH Verlag)

IOP Conf. Series: Journal of Physics: Conf. Series **1014** (2018) 012017 doi:10.1088/1742-6596/1014/1/012017

Modern hadron spectroscopy: a bridge between nuclear and particle physics.

A. P. Szczepaniak

Physics Department, Indiana University, Bloomington, IN 47405, USA,
Center for Exploration of Energy and Matter, Indiana University, Bloomington, IN 47403, USA,
Theory Center, Thomas Jefferson National Accelerator Facility, Newport News, VA 23606, USA.

E-mail: aszczepa@indiana.edu

Abstract. In this talk I discuss aspects of hadron physics, which soon are expected to shed new light on the fundamental QCD phenomena. In the analysis of hadron reactions and their propertieds I emphasize similarities to the nuclear many body problem.

1. Introduction

The vast majority of nuclear phenomena can be understood using protons and neutrons as elementary constituents and the nonrelativistic interactions among them. On the other hand, Quantum Chromodynamics (QCD), which is the underlying theory of nuclear forces, describes the relativistic quarks and gluons as the fundamental degrees of freedom. It is fascinating that the behavior and properties of the "elementary" hadrons, *i.e.* isolated bound states of quarks and gluons share many similarities with both, the low-energy nuclear phenomena as well as the high-energy QCD behavior.

2. Hadrons as a window for novel QCD phenomena

Confinement is the distinguishing feature of QCD [1]. It seems to forbid isolated elementary quarks and gluons to exist freely in nature. Understanding the origin of confinement is one of the fundamental questions in physics. Confinement is due to self-interactions between gluons, which are the mediators of the strong force between colored quarks and other gluons. The light, u and d quarks that make up the nuclei are O(100) times lighter than the proton and thus the relativistic interactions between gluons are in fact responsible for the bulk of the visible mass in the Universe. The same gluons that provide binding between quarks and produce the nucleon mass can also be active in the hadron spectrum. However, since gluons carry no charge other than color, gluons are invisible to external probes. For these reason, gluons remain the most mysterious particles of the Standard Model. Fortunately, their contribution to the hadron spectrum is expected to produce novel states of mater [2] and new experimental programs have just began searching for such matter, which is dominated by radiation and referred to as glueballs and hybrids. The latter contains both quarks and excited gluons.

Hadrons are the "simplest" bound states of quarks and gluons allowed by confinement and studies of individual hadrons and their interaction offer a unique window into this fundamental

IOP Conf. Series: Journal of Physics: Conf. Series **1014** (2018) 012017 doi:10.1088/1742-6596/1014/1/012017

phenomenon. Developments in particle accelerators and detection techniques have led to a new generation of experiments in hadron physics that are flourishing around the world. These facilities and experiments include existing laboratories such as BESIII in China, COMPASS and LHCb at CERN, RHIC in the US, VES in Russia, Belle in Japan and JLab in the US. PANDA in Germany, is in the construction phase, and future facilities under discussion include the Electron Ion Collider in the US. The new experiments at these facilities generate complicated data sets, which demand a qualitatively new level of sophistication in analysis never before achieved [3] In particular development of theoretical, phenomenological and computational underpinning of data analysis and interpretation is urgently needed to take full advantage of the information contained in the experimental data.

Since the early days of the hadron physics research, the quark model has provided the main template for "constructing" hadrons [4]. Over the years sophisticated extensions of the quark "shell model" of hadrons have been proposed [5, 6, 7, 8, 9]. While very successful in providing important insights into the nature of hadron interactions and their properties, such as the magnetic moment and charge distributions, the precise rooting of the quark model in QCD remains unsolved. In recent years, lattice gauge simulations have enabled first principle, numerical studies of QCD [10]. As a result, new insights into the hadron resonance spectrum have emerged. These have largely confirmed the naive quark model expectations that attributes the existence of the lightest hadrons to binding of quark-antiquark pairs for mesons and three quarks for baryons. The key question is how the almost massless quarks which, can in principle be copiously produced from gluon fusion become massive and that only a few of them are needed to build a hadron. A possible mechanism based on the mean filed approximation borrows from the techniques of Hartee-Fock-Bogolubov used in condensed matter and nuclear physics. In the QCD vacuum the Fermi-Dirac sea becomes perturbed when an external quark-antiquark are inserted and a condensate is formed. An excitation of this condensate, which is similar to a spin wave in a magnetic material can be identified with pions, which are the lightest of mesons [11]. The external quarks, because of the long range long range, confining interaction between their color charges and those of the individual quark in the medium acquire an effective mass. The mass is, however, finite and of the order of a few hundred MeV, because of the color neutrality of the vacuum. A similar mass generation mechanisms can apply to gluons. The main complication here arises from the strong gauge dependence. What a gluon is strongly depends on the choice of gauge. As it is in the case of nonrelativistic QED, QCD in the Coulomb gauge provides with the canonical relation between gluons as particles and gluons as fields. Unfortunately, QCD in the Coulomb gauge is highly nonlinear, which limits the reach of analytical approaches [12, 13]. From the numerical, lattice gauge simulations, it appears that gluons in the vacuum from complicated topological structures of monopoles and vortices [14]. In a popular confinement scenario, the condensate of chromo-magnetic monopoles screens the chromo-electric field emerging from the color charges of quarks and gluons, which instead of spreading out in a Coulomb-like pattern become collimated into flux tubes. This phenomenon is referred to as the dual Meissner effect (in the normal Meissner effect, in electric superconductor magnetic fields are screened by the freely moving electric charges). When this picture is considered in the Coulomb gauge, where besides the gluon condensate there are physical, particle-like gluons one finds that the latter, because of the interactions with the condensate behave effectively as massive axial vector particles, *i.e* with spin (J), parity (P) and charge conjugation (C) quantum numbers of $J^{PC} = 1^{+-}$ [15, 16]. When such a quasi-gluon is combined with a quark antiquark pair, that has quantum numbers of $J^{PC} = 0^{-+}$ in the spin zero state and 1^{--} in the spin-one state one obtains four states with the J^{PC} quantum numbers, 1^{--} and $(0,1,2)^{-+}$ [17]. These states are interpreted describing the four lightest hybrids mesons, that include the so called exotic hybrid, which is the one with $J^{PC} = 1^{-+}$. It is exotic, because unlike the other combination of the J, P and C quantum numbers, the exotic combination cannot be obtained from combining quark and antiquark quantum numbers alone. Therefore, a resonance with exotic quantum numbers is being extensively searched for a smoking gun indication of gluon excitation in the hadron spectrum. The patter of hybrid mesons that emerge in Coulomb gauge has recently been confirmed by lattice simulations [18]

Identification of exotic hadrons in the spectrum requires sophisticated methods of partial wave and reaction amplitude analysis. This is because in the mass region were the new hadrons are expected there are several open channels that contribute to the non-resonance backgrounds complicating extraction of the signal. When treated as complex functions of the kinematical variables, reaction amplitudes have singularities that originate from presence of resonances. Thus, to properly extract resonance parameters it is necessary to develop analytical reaction amplitudes.

Several new experiments at the Jefferson Lab that began operation this year will be searching for the hybrid meson and baryon resonances. The experiments use the 12 GeV photon beam from the energy upgraded CEBAF accelerator that are scattered peripherally on the hydrogen target (for meson resonance extraction) or centrally, for studies of the baryon spectrum. In case of (point-like) photon-induced reactions kinematical separation of the beam and target fragmentation is highly nontrivial and requires deep understanding of the underlying production process. For example, microscopically, the peripheral photo-production involves scattering of photons on the virtual meson cloud surrounding the proton target and should be described with the framework of the Regge theory. A particularly interesting aspect of this reaction involves production of the lightest pions [19, 20]. Specifically, through studies of the photon polarization asymmetries, one can isolate production via exchange of natural. (ρ, ω) vs unnatural parity (b, h) reggeons. Recent theoretical predictions by the JPAC collaboration indicating dominance of the natural component have been conferment by in the first measurement performed with GlueX detector [21]. Confirmation of the theoretical production models by these early measurements opens path for rigorous analysis of the data and potential discovery of the new QCD phenomenal associated with gluon excitations.

3. Summary

New advancements in experimental techniques, numerical and analytical approaches to QCD are expected to provide key insights into the fundamental aspects of QCD, such as confinement and mass generation. These advancements will help to further integrate the hadron and nuclear physics manifestations of the strong interactions.

Acknowledgments

This work is supported in part by the U.S. Department of Energy, Office of Science, Office of Nuclear Physics under contract DE-AC05-06OR23177, DE-FG0287ER40365.

References

[1] Gross D in *Proceedings of the 2000 International Superstrings Conference University of Michigan, Ann Arbor.*
[2] Briceno R A *et al.* 2016 *Chin. Phys. C* **40** 042001
[3] Battaglieri M *et al.* 2015 *Acta Phys. Polon. B* **46** 257
[4] Shepherd M R, Dudek J J and Mitchell R E 2016 *Nature* **534** 487
[5] Godfrey S and Isgur N 1985 *Phys. Rev. D* **32** 189
[6] Capstick S and Isgur N *Phys. Rev. D* **34** 2809
[7] Roberts C D and Williams A G 1994 *Prog. Part. Nucl. Phys.* **33** 477
[8] Bijker R and E. Santopinto E 2009 *Phys. Rev. C* **80** 065210
[9] Eichmann G *et al.* 2016 *Prog. Part. Nucl. Phys.* **91** 1
[10] Briceno R A, Dudek J J and Young R D arXiv:1706.06223 [hep-lat].
[11] Szczepaniak A P and Swanson E S 2001 *Phys. Rev. Lett.* **87** 072001
[12] Christ N H and Lee T D *Phys. Rev. D* **22** 939
[13] Gribov V N 1978 *Nucl. Phys. B* **139** 1

[14] Greensite J 2003 *Prog. Part. Nucl. Phys.* **51** 1
[15] Szczepaniak A P and Krupinski P 2006 *Phys. Rev. D* **73** 116002
[16] Guo P, Szczepaniak A P, Galata G, Vassallo A and Santopinto E 2008 *Phys. Rev. D* **77** 056005
[17] Guo P, Szczepaniak A P, Galata G, Vassallo A and Santopinto E 2008 *Phys. Rev. D* **78** 056003
[18] Dudek J J, Edwards R G, Peardon M J, Richards D G and Thomas C E 2010 *Phys. Rev. D* **82** 034508
[19] Mathieu V *et al.* 2017 *Phys. Lett. B* **774** 362
[20] Nys J *et al.* [JPAC Collaboration], arXiv:1710.09394 [hep-ph].
[21] Ghoul H Al *et al.* [GlueX Collaboration] 2017 *Phys. Rev. C* **95** 042201

A study on the dependence of nuclear viscosity on temperature

E Vardaci[1], A Di Nitto[1,2], P N Nadtochy[3], G La Rana[1], M Cinausero[4], G Prete[4], N Gelli[5], M Ashaduzzaman[1], F Davide[1], A Pulcini[1], D Quero[1], E M Kozulin[6], G N Knyazheva[6] and I M Itkis[6]

[1]Dipartimento di Fisica "E. Pancini", University of Naples "Federico II" and INFN, I-80126 Napoli, Italy.
[2]Johannes Gutenberg- Universität Mainz Institut für Kernchemie, Mainz, Germany.
[3]Omsk State Technical University, Mira prospekt 11, 44050 Omsk, Russia.
[4]INFN - Laboratori Nazionali di Legnaro, Legnaro (Padova), Italy.
[5]INFN - Firenze, Firenze, Italy.
[6]Flerov Laboratory of Nuclear Reactions, Joint Institute for Nuclear Research, RU-141980 Dubna, Moscow Region, Russia.

E-mail: vardaci@na.infn.it

Abstract. Nuclear viscosity is an irreplaceable ingredient of nuclear fission collective dynamical models. It drives the exchange of energy between the collective variables and the thermal bath of single particle degrees of freedom. Its dependence on the shape and temperature is a matter of controversy. By using systems of intermediate fissility we have demonstrated in a recent study that the viscosity parameters is larger for compact shapes, and decreases for larger deformations of the fissioning system, at variance with the conclusions of the statistical model modified to include empirically viscosity and time scales. In this contribution we propose an experimental scenario to highlight the possible dependence of the viscosity from the temperature.

1. Introduction

It is well known that the process of nuclear fission is an ideal laboratory for probing nuclear dissipation [1, 2]. Experimental evidences on how nuclear viscosity affects the compound fission process, as well as the competition with the fusion-evaporation process, has come from the observation of the evolution of many observables such as prescission multiplicities of neutrons, charged particles and γ-rays with the excitation energy. Broadly speaking, these quantities appear to be significantly larger than those predicted by the statistical model. The main important consequence of fission viscosity is to slow down the collective motion toward the separation in two fragments with the implication that prescission radiations are more likely to escape. The bulk results are an increase of their multiplicity and a higher probability of producing evaporation residues. Applications of the statistical model which include, with a specific ansatz, the effects of viscosity give rather controversial estimates of the time scale and strength of viscosity [3-9].

Indeed, a dynamical picture of the fission process affected by viscosity cannot be effectively included in the statistical model. This matter has been discussed in some detail in [9]. Dynamical models can include dissipation effects due to viscosity and fluctuations ab-initio [10-14]. In the past years we have developed a 3D dynamical model based on the Langevin approach to fission to calculate the decay of light particles from the fissioning nucleus [15, 14]. We have applied this model to our extensive set of data gathered on systems of intermediate fissility [8, 15] and have extracted a more convincing picture of the nature of dissipation and its dependence on the evolving shape of the nucleus. The two main results are summarized in figure 1. The viscosity

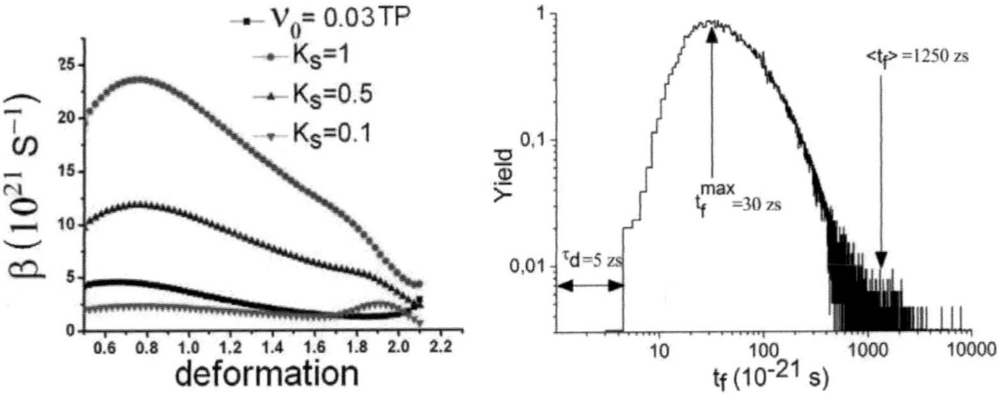

Figure 1. Main results of the 3D calculations for the system $^{32}S+^{100}Mo$ at 200 MeV. Left:Reduced viscosity parameter vs. deformation of the compound nucleus. K_s is the strength of the one-body dissipation. Full strength is for $K_s = 1$. The black solid line is the functional dependency expected in the case of the two-body dissipation. Right:Calculated distribution of the fission times.

parameter is found to be strongly dependent on the shape of the nucleus undergoing fission, being larger for compact shapes (around the equilibrium configuration) and smaller at the more elongated shapes (shapes around the saddle point and more deformed). In particular, our set of data agrees with the one-body nature of dissipation represented by $K_s = 1$ (full strength). One unprecedented outcome of our model is the calculation of the time distribution of the fission events (figure 1, right). This distribution typically extends over 3 orders of magnitude and takes into account multiple chance fission. The time delay parameter widely used in the statistical approach does not correspond to any of the characteristic times of this distribution. This result confirms the inadequacy of the statistical model approach to nuclear dissipation. The wide extension of the time distribution found also explains why different time scales are extracted within statistical model implementations of nuclear viscosity when different probes are used. The results of this work [15] have further implications on the dynamical evolution of composite systems in other non-compound processes like quasifission [16].

2. How to probe a temperature-dependent viscosity

In the above calculations nuclear viscosity is assumed independent from the temperature. The nuclear mechanism of dissipation may indeed depend on the temperature. A pathway to explore such a dependence was undertaken by the following observations. In many experimental studies on heavy systems it is commonly found that if the temperature of the compound nucleus T_{CN} is raised, the temperature at scission point T_{SC} remains almost constant [1, 17, 18]. This supports the picture in which fission occurs after that the composite nucleus has cooled down until it has reached an excitation energy of roughly 50-60 MeV, independently on the excitation energy of the initial compound nucleus. This is shown in figure 2 (left) [17].

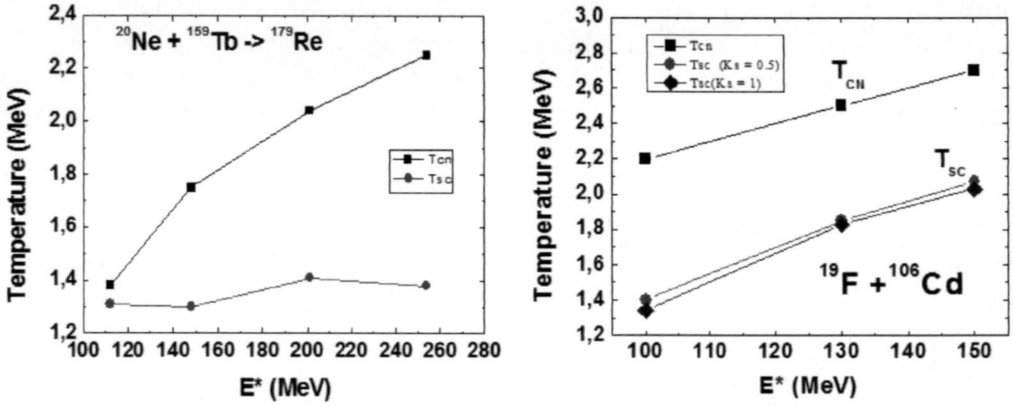

Figure 2. Left: Pre- and post-scission temperatures vs. initial excitation energy of the compound nucleus [17]. Right: Pre- and post-scission temperatures vs. initial excitation energy of the compound nucleus in the case of an intermediate mass system calculated with a dynamical model [14]. $K_s = 1$ is for full one-body dissipation.

If a study on the dependence of nuclear viscosity on the temperature is performed with this picture in mind (for instance by measuring the excitation function of some observables of the kind above), the measurable effects on the observables, which are deduced from the deviations from the model used, are averaged over an interval of temperatures $\Delta T = T_{CN}$-T_{SC} which grows with the bombarding energy. In other words, as the interval of temperatures T_{CN}-T_{SC} increases, the observable effects are integrated over a larger and larger interval of temperatures. A possible dependence on the temperature will be therefore hidden. Consequently, heavy systems, like the one in figure 2 (left), are not the ideal frame for the study of the dependence of the viscosity parameter from the temperature.

Systems of intermediate fissility have instead a completely different behavior. With our model [14] we calculated T_{SC} for several systems. In figure 2 (right) we show the dependence of T_{CN} and T_{SC} from the initial excitation energy of the compound nucleus for the case of the system ^{19}F+^{106}Cd. The interesting feature of these systems is that T_{SC} also grows with the excitation energy of the initial compound nucleus, a behavior at variance with that found for heavier systems. This is due to the delicate balance between the time scale of the fission process and the binding energy of the pre-scission particles: the neutron pre-scission emission, which is mainly responsible for the cooling in the pre-saddle region, is suppressed by the higher binding energy and stronger competition with charged particles. As a result, the two fission fragments retain most of the initial excitation energy.

3. Experiment and results

An experiment was designed to probe the dependence of the nuclear viscosity on temperature within the picture above, keeping in mind that in case we are unable to reproduce a large set of data with our model for different excitation energies, a dependence on the temperature must be taken into account considering the features of the deviation.

In the present contribution we report on the preliminary results of an experiment performed at Laboratori Nazionali di Legnaro using the 4π array 8πLP. For the system ^{19}F+^{106}Cd, at two lab energies 175 and 195 MeV, we have measured most of the relevant quantities in the evaporation residues (ER) and fission channel (FF) channels: proton and α-particle energy spectra and

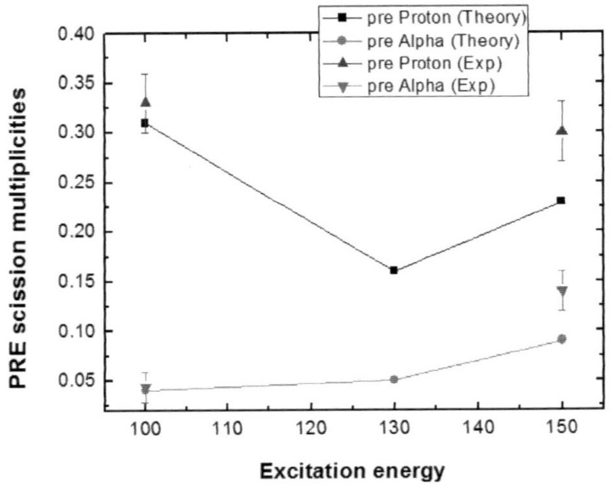

Figure 3. Pre-scission proton and α particles multiplicities measured for the system $^{19}F+^{106}Cd$ compared to the full one-body dissipation dynamical calculations assuming no temperature dependence.

multiplicities, ER and FF cross sections, as well as mass and total-kinetic-energy distributions of fission fragments. The extraction procedure of the proton and α-particle multiplicities from the energy spectra measured in coincidence with ER and FF is the same as in Refs.[8, 15]. Modeling of the fission process has been performed using again the same model as above.

In figure 3 we show the pre-scission protons and α-particle multiplicities compared to the model results. We observe that at lower excitation energy, the model reproduces the experimental data, whereas at the higher excitation energy the experimental multiplicities are underestimated. This kind of trend suggests that a stronger viscosity would be necessary at higher temperature.

4. Conclusions

The analysis presented here remarks how promising is the dynamical approach in describing both fission and ER channel within the same model. In this contribution we have proposed an alternative scenario where the study of the dependence of the nuclear viscosity from the temperature seems more suitable. From the preliminary results, it seems that nuclear viscosity is slightly dependent on the temperature. However, this conclusion would indeed need further checks because the comparison with the model is so far limited to only two observables. A more stringent conclusion can be drawn after analyzing all together also the other observables still to be extracted from the experimental data.

References

[1] Hilscher D and Rossner H 1992, *Ann. Phys. (France)* **17** 471 and references therein
[2] Paul P and Thoennesen M 1994 *Ann. Rev. Nucl. Part. Sci.* **44** 65 and references therein
[3] Back B B *et al* 1999 *Phys. Rev.* C **60** 044602
[4] Chatterijee A *et al* 1995 *Phys. Rev.* C **52** 3167
[5] Dioszegi I *et al* 2000 *Phys. Rev.* C **61** 024613
[6] La Rana G *et al* 2003 *Eur. Phys. J.* A **16** 199
[7] Vardaci E *et al* 2010 *Eur. Phys. J.* A **43** 127
[8] Di Nitto A *et al* 2011 *Eur. Phys. J.* A **47** 83

IOP Conf. Series: Journal of Physics: Conf. Series **1014** (2018) 012018 doi:10.1088/1742-6596/1014/1/012018

[9] Vardaci E *et al* 2013 *EPJ Web Conf.* **62** 07004
[10] Wada T *et al* 1993 *Phys. Rev. Lett.* **70** 3538
[11] Frobrich P and Gontchar I I 1998 *Phys. Rep.* **292** 131 and references therein
[12] Karpov A V *et al* 2001 *Phys. Rev.* C **63** 54610
[13] Nadtochy P N *et al* 2002 *Phys. Rev.* C **65** 64615
[14] Nadtochy P N *et al* 2010 *Phys. Lett.* B **685** 258
[15] Vardaci E *et al* 2015 *Phys. Rev.* C **92** 034610
[16] Itkis M G, Vardaci E, Itkis I M, Knyazheva G N and Kozulin E M 2015 *Nucl. Phys.* A **944** 204 and references therein
[17] Cabrera J *et al* 2003 *Phys. Rev.* C **68** 34613
[18] Mordhorst E *et al* 1991 *Phys. Rev.* C **43** 716

IOP Conf. Series: Journal of Physics: Conf. Series **1014** (2018) 012019 doi:10.1088/1742-6596/1014/1/012019

The effects of collision orientation and energy dependence in multinucleon transfer reactions

Jingjing Li[1,2], Cheng Li[1,2], Peiwei Wen[3] and Feng-Shou Zhang[1,2,4]

E-mail: fszhang@bnu.edu.cn

[1]The Key Laboratory of Beam Technology and Material Modification of Ministry of Education, College of Nuclear Science and Technology, Beijing Normal University, Beijing 100875, China

[2]Beijing Radiation Center, Beijing 100875, China

[3]China Institute of Atomic Energy, Beijing 102413, China

[4]Center of Theoretical Nuclear Physics, National Laboratory of Heavy Ion Accelerator of Lanzhou, Lanzhou 730000, China

Abstract. Multinucleon transfer (MNT) reaction ^{136}Xe+^{208}Pb near Coulomb barrier energies are investigated within the dinuclear system (DNS) model. It is found that the collision orientation has an important influence on the mass distributions attributed to the depth of pocket in the driving potential. The calculation results of the isotopic production show that the energy dependence in neutron-deficient side is more sensitive than that in neutron-rich side. The production of the $N = 126$ isotones are calculated by GRAZING model, DNS+GEMINI model, and ImQMD+GEMINI model, respectively. It demonstrates that MNT reaction is a promising way to produce neutron-rich isotopes in the region of the neutron shell closure $N = 126$.

1. Introduction

The synthesis of unknown heavy neutron-rich nuclei in MNT reactions has becomes the focus of attention in nuclear physics recently [1–6]. The discovery of these nuclei is particularly important in understanding the r-process in nuclear astrophysics. Projectile fragmentation is an important method to produce new neutron-rich isotopes in the atomic number range of $Z < 92$ in the past several decades, but the production cross sections is quite small [7]. In recent years, experimental research on MNT reactions have been performed at available facilities by different groups. For example, reaction of ^{136}Xe+^{208}Pb was carried out at the Flerov Laboratory of Nuclear Reactions in Dubna in 2012 and at the Gammasphere facility of Argonne National Laboratory in 2015 to produce the neutron-rich $N = 126$ nuclei [8, 9]. Meanwhile, reaction of ^{136}Xe+^{198}Pt at the energy above the Coulomb barrier was performed by Watanabe group at GANIL in Caen [10]. It demonstrates that MNT reactions have obvious advantages in the production of new heavy neutron-rich isotopes around $N = 126$ shell [3].

To understand the MNT progress in heavy-ion collisions, several theoretical models have been proposed. The phenomenological models such as the DNS model [6, 11] and GRAZING code [12, 13] have been developed and used by different groups for predicting the production cross sections of heavy neutron-rich nuclei. In addition, some microscopic dynamics models such as the time-dependent Hartree-Fock (TDHF) model [14] and improved quantum molecular dynamics (ImQMD) model [4, 5] have been successful used to describe the nucleon transport and energy dissipation during the heavy-ion collisions. In this work, we investigate the influence of the collision orientation and energy dependence in MNT reaction of ^{136}Xe+^{208}Pb within the framework of the DNS model.

The article is organized as follows. In section 2, a brief description of the theoretical models. The results and discussion are presented in section 3, and conclusions are given in section 4.

2. Brief description of the DNS model

In the DNS model, the evolution of the colliding partner is a diffusion process along mass asymmetry and relative distance degrees of freedom. The potential energy surface (PES) of the DNS is given by

$$\begin{aligned} & U(Z_1, N_1, Z_2, N_2, R, \beta_1, \beta_2, \theta_1, \theta_2, J) \\ = \ & U_{LD}(Z_1, N_1) + U_{LD}(Z_2, N_2) - U_{LD}(Z, N) \\ - \ & V_{rot}^{CN}(J) + V_{CN}(Z_1, N_1, Z_2, N_2, R, \beta_1, \beta_2, \theta_1, \theta_2, J). \end{aligned} \quad (1)$$

Here $U_{LD}(Z_1, N_1)$, $U_{LD}(Z_2, N_2)$ and $U_{LD}(Z, N)$ are the binding energies of two deformed nuclei and CN, respectively. $V_{rot}^{CN}(J)$ is the rotation energy of the CN. $V_{CN}(Z_1, N_1, Z_2, N_2, R, \beta_1, \beta_2, \theta_1, \theta_2, J)$ is the interaction potential of two fragments. Detail description of V_{CN} see Ref. [15]. Within the PES, the distribution probability of a fragment is obtained by solving the follow master equation [6, 11, 15]:

$$\begin{aligned} & \frac{dP(Z_1, N_1, E_1, t)}{dt} \\ = \ & \sum_{Z_1'} W_{Z_1,N_1;Z_1',N_1}(t)[d_{Z_1,N_1} P(Z_1', N_1, E_1', t) - d_{Z_1',N_1} P(Z_1, N_1, E_1, t)] \\ + \ & \sum_{N_1'} W_{Z_1,N_1;Z_1,N_1'}(t)[d_{Z_1,N_1} P(Z_1, N_1', E_1', t) - d_{Z_1,N_1'} P(Z_1, N_1, E_1, t)] \\ - \ & [\Lambda_{qf}(\Theta(t)) + \Lambda_{fis}(\Theta(t))] P(Z_1, N_1, E_1, t). \end{aligned} \quad (2)$$

The production cross section of a primary fragment with charge number Z and mass number A are calculated as follows:

$$\begin{aligned} \sigma_{tr}^{DNS}(Z, A, E) \ = \ & \frac{\pi \hbar^2}{2\mu E} \sum_l (2l + 1) \\ \times \ & P_c(E, l) P_{tr}^{DNS}(Z, A, E), \end{aligned} \quad (3)$$

where μ is the reduced mass of the system and $E_{\mathrm{c.m.}}$ is the incident energy at the center of mass frame. P_c is capture probability and $P_{tr}^{DNS}(Z, A, E)$ is the production probability of a fragment. The code GEMINI is used to treat the excited primary fragments in the subsequent de-excitation processes.

3. Results and Discussion

In Fig. 1, the driving potential as a function of mass number in reaction ^{136}Xe+^{208}Pb at different collision orientations are presented. In this work, the orientation of the light fragment θ_1 is set to be 0°. It is shown that the collision orientation of two fragments has an important influence on the depth of the pocket in the driving potential. One can see that the driving potential increases with the increasing of the collision orientation θ_2. In fact, the driving potential has a very important influence on the transfer of the nucleons.

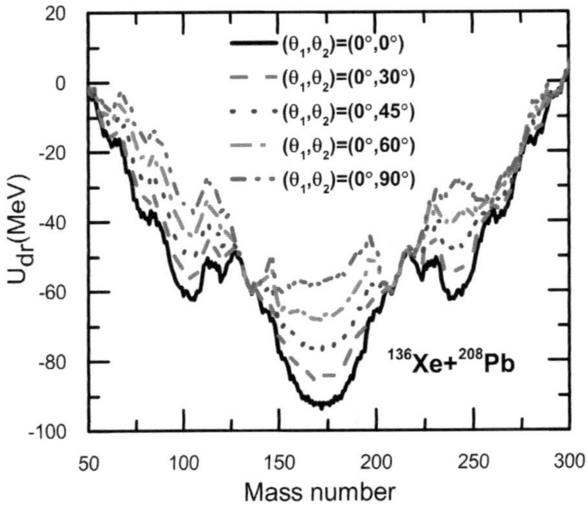

Figure 1. The driving potential as a function of mass number in reaction ^{136}Xe+^{208}Pb at different collision orientations.

Mass distributions of primary and secondary fragments in reaction ^{136}Xe+^{208}Pb at $E_{c.m.}$ = 450 MeV at different collision orientations are shown in Fig. 2. The solid and dashed lines denoted the mass distributions of primary and secondary fragments, respectively. The black, blue, olive indicate the different collision orientation. The red circles represent the available experimental data from Argonne National Laboratory [9]. From Fig. 2, one can see that the mass distributions of both primary and secondary fragments are influenced by the collision orientations. Note that the yields of the fragments in the region of 150⩽A⩽190 decrease with the increase of the collision orientation of θ_2 due to the higher potential in the driving potential as shown in Fig. 1.

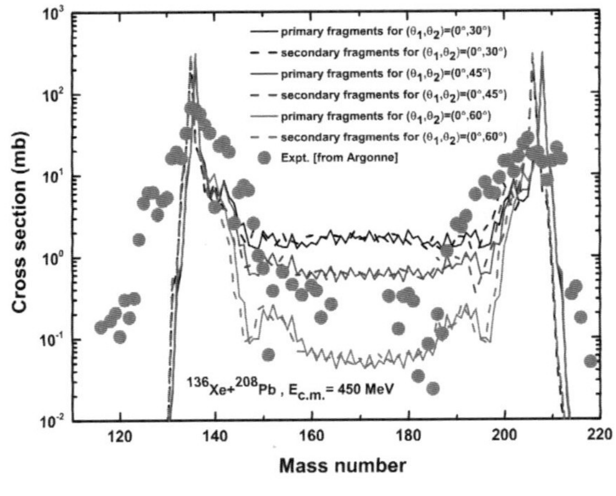

Figure 2. Orientation effects on the mass distributions of primary and secondary fragments in reaction ^{136}Xe+^{208}Pb at $E_{c.m.}$ = 450 MeV and compared with the available experimental data from Argonne [9].

Fig. 3 shows the final isotopic production cross sections from Pt to At for reaction

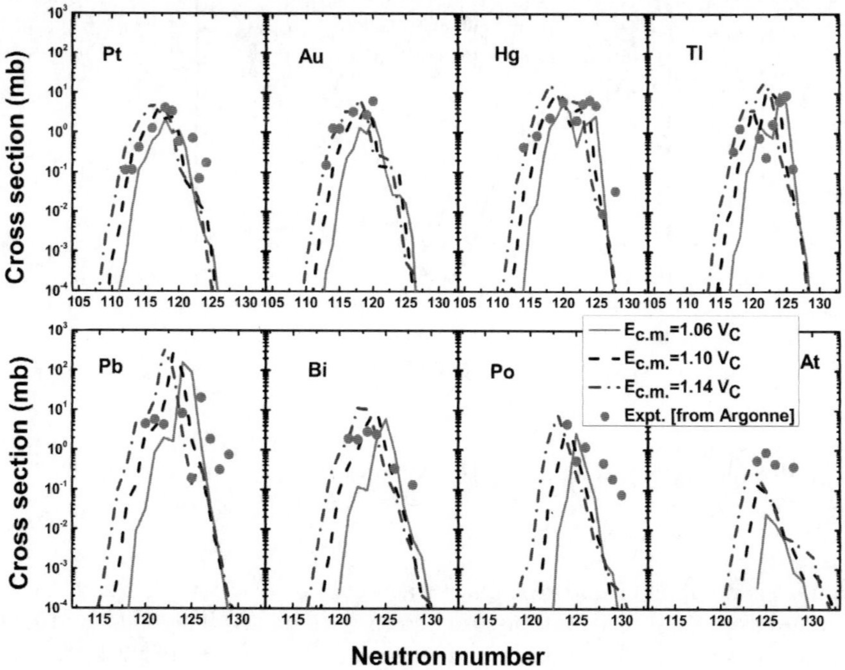

Figure 3. The isotopic production cross sections from Pt to At for reaction $^{136}Xe+^{208}Pb$ at $E_{c.m.} = 1.06, 1.10$, and $1.14\ V_C$. Here V_C is the Coulomb barrier. The experimental data (solid circle) are from Ref. [9].

$^{136}Xe+^{208}Pb$ at different energies. It is notable that the energy dependence in neutron-deficient side is more sensitive than that in neutron-rich side. One important reason is that higher beam energy leads to the higher local excitation energy which makes the primary fragments easy to obtain more excitation energy. Then the excited fragments in neutron-rich side will de-excite by evaporating more neutrons. In consequence, the final isotopic distributions shift from neutron-rich side to neutron-deficient side. In addition, higher incident energy can improve the transfer probability of nucleons within the same PES. However, higher beam energy leads to the higher local excitation energy which leads to the smaller survival probability of primary fragments, especially for extremely neutron-rich nuclei. It is obvious that the mass distributions near the projectile-like fragments reduce rapidly. This is because there is a steeply rise in the driving potential when ^{136}Xe transfer nucleons to ^{208}Pb.

In Fig. 4, the production cross sections of the $N = 126$ isotones in reaction $^{136}Xe+^{208}Pb$ at $E_{c.m.} = 450$ MeV are calculated by GRAZING model, DNS+GEMINI model, and ImQMD+GEMINI model, respectively. A comparison of the theoretical calculations with the experimental measurements [9] is also shown. One can see that GRAZING model is suitable for the transfer of just a few nucleons. ImQMD+GEMINI model is successful in description of a large number of nucleons transfer. For DNS+GEMINI model, the calculation results are consistent with the experimental data near the peak. The neutron-rich nuclei ^{204}Pt, ^{205}Au, ^{206}Hg, ^{207}Tl can be produced with quite high cross sections in this reaction. It demonstrates

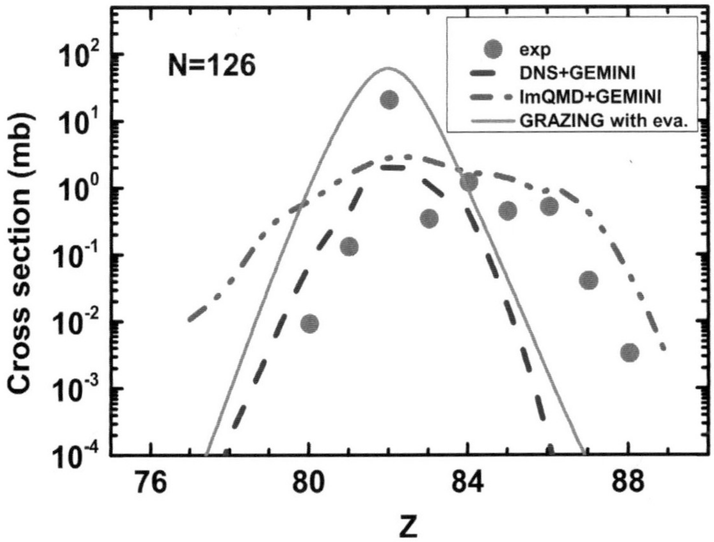

Figure 4. The production cross sections of the $N = 126$ isotones as a function of the atomic number in reaction ^{136}Xe+^{208}Pb at $E_{c.m.} = 450$ MeV. The solid, dashed, and dash-dotted lines indicate the calculation results from GRAZING model, DNS+GEMINI model, and ImQMD+GEMINI model, respectively. The solid circles denote the available experimental data from Argonne [9]

.

that the MNT reaction is a promising way to produce neutron-rich isotopes in the region of the neutron shell closure $N = 126$.

4. Conclusions

The influence of the collision orientation and energy dependence in MNT reaction of ^{136}Xe+^{208}Pb is studied by using the DNS model. The collision orientation has an important influence on the mass distributions due to the depth of the pocket in the driving potential. The isotopic production cross sections are predicted at $E_{c.m.} = 1.06$, 1.10, and 1.14 V_C and compared with the available experimental data. We find that the energy dependence in neutron-deficient side is more sensitive than that in neutron-rich side. The MNT reaction is regarded as a promising method to produce neutron-rich isotopes in the region of the neutron shell closure $N = 126$.

Acknowledgments

This work was supported by the National Natural Science Foundation of China under Grants No. 11635003, No. 11025524 and No. 11161130520; the National Basic Research Program of China under Grant No. 2010CB832903; the European Commission's 7th Framework Programme (Fp7-PEOPLE-2010-IRSES) under Grant Agreement Project No. 269131.

[1] Devaraja H M *et al.* 2015 *Phys. Lett. B* **748** 199
[2] Comas V F *et al.* 2013 *Eur. Phys. J. A* **49** 112
[3] Beliuskina O *et al.* 2014 *Eur. Phys. J. A* **50** 161
[4] Li C *et al.* 2018 *Phys. Lett. B* **776** 278
[5] Zhao K *et al.* 2016 *Phys. Rev. C* **94** 024601
[6] Zhu L *et al.* 2017 *Phys. Lett. B* **767** 437
[7] Kurcewicz J *et al.* 2012 *Phys. Lett. B* **717** 371
[8] Kozulin E M *et al.* 2012 *Phys. Rev. C* **86** 044611
[9] Barrett J S *et al.* 2015 *Phys. Rev. C* **91** 064615
[10] Watanabe Y X *et al.* 2015 *Phys. Rev. Lett.* **115** 172503
[11] Adamian G G *et al.* 1997 *Nucl. Phys. A* **618** 176
[12] Winther A *Nucl. Phys. A* 1994 **572** 191
[13] Yanez R and Loveland W 2015 *Phys. Rev. C* **91** 044608
[14] Sekizawa K and Yabana K 2014 *Phys. Rev. C* **90** 064614
[15] Feng Z Q *et al.* 2009 *Phys. Rev. C* **80** 057601

Institute of Physics
Dirac House, Temple Back
Bristol BS1 6BE UK

ISSN: 1742-6588
ISBN 978-1-5108-6290-6

CURRAN ASSOCIATES INC.
proceedings.com

18th International Workshop on Advanced Computing and Analysis Techniques in Physics Research (ACAT 2017)

AI Reloaded

Journal of Physics: Conference Series
Volume 1085

Seattle, Washington, USA
21 – 25 August 2017

IOP
Institute of Physics